ELECTRICAL AND OPTICAL PROPERTIES OF III–V SEMICONDUCTORS

ELEKTRICHESKIE I OPTICHESKIE SVOISTVA POLUPROVODNIKOV $A^{III}B^{V}$

ЭЛЕКТРИЧЕСКИЕ И ОПТИЧЕСКИЕ СВОЙСТВА ПОЛУПРОВОДНИКОВ $A^{III}B^{V}$

The Lebedev Physics Institute Series

Editors: Academicians D. V. Skobel'tsyn and N. G. Basov

P. N. Lebedev Physics Institute, Academy of Sciences of the USSR

Recent Volumes in this Series

In preparation

Proceedings (Trudy) of the P. N. Lebedev Physics Institute

Volume 89

Electrical and Optical Properties of III-V Semiconductors

Edited by
N. G. Basov
P. N. Lebedev Physics Institute
Academy of Sciences of the USSR
Moscow, USSR

Translated from Russian by
Albin Tybulewicz
Editor, *Soviet Physics - Semiconductors*

CONSULTANTS BUREAU
NEW YORK AND LONDON

Library of Congress Cataloging in Publication Data

Main entry under title:

Electrical and optical properties of III-V semiconductors.
(Proceedings (Trudy) of the P.N. Lebedev Physics Institute; v. 89)
Translation of Elektricheskie i opticheskie svoĭstva poluprovodnikov $A^{III} B^{V}$.
Includes bibliographies and index.
1. Semiconductors. 2. Semiconductors–Optical properties. 3. Gallium arsenide crystals. 4. Indium antimonide crystals. I. Basov, Nikolaĭ Gennadievich, 1922-
II. Series: Akademiia nauk SSSR. Fizicheskiĭ institut. Proceedings; v. 89.
QC1.A4114. vol. 89 [QC611] 530'.08s [537.6'22]
ISBN 0-306-10944-1 77-26132

The original Russian text was published by Nauka Press in Moscow in 1976 for the Academy of Sciences of the USSR as Volume 89 of the Proceedings of the P. N. Lebedev Physics Institute. This translation is published under an agreement with the Copyright Agency of the USSR (VAAP).

A Division of Plenum Publishing Corporation
227 West 17th Street, New York, N.Y. 10011

Printed in the United States of America

PREFACE

This volume consists of two long papers. The first reports in detail an investigation of the electrical conductivity of moderately doped compensated gallium arsenide. The second describes an investigation of spontaneous and coherent emission of radiation from pure and doped indium antimonide as a result of the double injection of carriers.

Both papers are aimed at a wide range of scientists, researchers, engineers, and chemical technologists investigating and fabricating semiconducting materials used in optoelectronics.

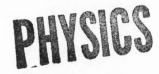

CONTENTS

LOCALIZATION OF ELECTRONS IN COMPENSATED GALLIUM ARSENIDE*

I. D. Voronova

An investigation was made of the electrical conductivity of moderately doped compensated gallium arsenide. The experimental data on the temperature dependence of the free-electron density were used in constructing an energy scheme which includes a band of localized states adjoining the bottom of the free-electron (conduction) band. Variation of the degree of compensation in a series of samples gave rise to an effect analogous to the Mott transition in uncompensated semiconductors. This "induced Mott transition" was observed at low temperatures in samples in which some electrons were localized when a certain nonequilibrium electron density was established by ruby laser illumination. Heating of the conduction electrons by a strong electric field showed that their effective mass was close to the effective mass of free electrons in pure gallium arsenide.

INTRODUCTION

The growth of semiconductor physics is characterized by intensive studies of doped materials because a practically unlimited range of physical properties which can be obtained by doping enhances greatly the chances of practical applications.

The impurity concentrations in semiconducting materials can be used to divide them into three groups. If the impurity concentration N_i is such that $N_i a_0^3 \gg 1$, where a_0 is the Bohr radius of an impurity atom, a semiconductor is called heavily doped. If $N_i a_0^3 \ll 1$, a semiconductor is known as lightly doped.

Moderate doping, i.e., the case when $N_i a_0^3 \sim 1$, is least investigated and is difficult to describe. Therefore, further studies are needed before moderately doped semiconductors can find practical applications.

The properties of moderately doped semiconductors are governed by the phenomena associated with the Mott–Hubbard transitions which occur in doped and uncompensated semiconductors at a certain critical impurity concentration [1]. These transitions alter the energy structure and the nature of conduction; studies of these transitions touch upon general aspects of solid state physics such as collective electron effects, phase transitions, changes in the crystal lattice, etc.

* Based on a thesis submitted for the degree of Candidate of Physicomathematical Sciences, defended at the P. N. Lebedev Physics Institute, Academy of Sciences of the USSR, Moscow. The work was carried out under the direction of Academician B. M. Vul and Candidate of Physicomathematical Sciences É. I. Zavaritskaya.

Similar transitions may occur in semiconductors with increasing degree of compensation, but the range of the phenomena in strongly compensated semiconductors is even wider. In addition to metal–insulator transitions, due to the electron correlation (formation of a Wigner electron crystal [2]), we can have electron transitions to localized states which appear because of fluctuations in the distribution of electrically active impurities. Thus, compensated semiconductors are a special case of a wide class of disordered materials, which also includes heavily doped, amorphous, and liquid semiconductors.

It is nowadays assumed that all such materials should have certain fundamental properties in common and studies of these materials are increasing rapidly in importance [3]. It has been established theoretically that the distribution of the density of states in compensated semiconductors has an energy limit separating localized and nonlocalized states [4, 5]. Mott assumes that an increase in the degree of compensation should result in a descent of the Fermi level at $T = 0$ below this boundary and then in the limit $T \rightarrow 0$ the conduction process should be of the hopping type. The activation energy of such hopping conduction should decrease as a result of cooling so that $\rho \propto T^{-1/4}$, where ρ is the resistivity. In fact, many experimental investigations of the conductivity of various compensated semiconductors have revealed a common feature, which is a weakening of the temperature dependence of the conductivity. However, at very low temperatures, this dependence becomes even weaker than predicted from the Mott law for the hopping conduction. The failure to explain this effect [6] and other data confirming the hopping nature of the conduction has left unresolved the question of the nature of conduction in compensated semiconductors. Moreover, little work has been done on the energy structure of these materials.

This situation makes it essential to carry out further experimental and theoretical investigations of these materials, particularly at low temperatures.

The investigation reported below was carried out on gallium arsenide, which is becoming of increasing importance in modern semiconductor electronics. Moreover, a high degree of compensation can be achieved in gallium arsenide [7], which makes it possible to carry out studies in a wide range of degrees of compensation.

CHAPTER I

REVIEW OF THE LITERATURE. LOCALIZATION OF ELECTRONS IN DOPED AND COMPENSATED SEMICONDUCTORS

Lightly and heavily doped semiconductors differ, as their designations indicate, by the numbers of impurities they contain. Quantitative differences in the limiting cases of light and heavy doping result in such considerable qualitative differences that, at $T = 0$, the lightly doped materials are insulators and the heavily doped are metals.

In the light doping case, the wave functions of electrons of neighboring impurity atoms do not overlap and are practically identical with the wave functions of free atoms. In this case, the impurity levels are separated from the conduction band by an energy gap and, if the temperature is sufficiently low and there is no compensation, the impurity atoms are practically nonionized.

At $T = 0$, a material of this kind is an insulator [8, 9].

When the impurity concentration is increased, the wave functions of the impurity centers begin to overlap and the impurity levels form a band. The energy gap between this impurity band and the conduction band decreases with increasing impurity concentration and at some concentration the two bands merge. Therefore, even in the limit $T \to 0$, the conduction band contains free electrons and a semiconductor can be regarded as a metal because it has a finite conductivity.

The continuous changes which occur in the energy band structure as a result of an increase in the concentration of shallow donors or acceptors were described back in the fifties [10, 11]. In some of the papers published since 1961, Mott was the first to show that the changes considered here could not occur continuously [4, 12]. He assumed that there should be a critical impurity concentration at which the properties of a semiconductor change abruptly. At this concentration, there is an insulator–metal transition known as the Mott transition.

§ 1. Mott Transition in an Ideal Crystal Lattice

The conclusion that there should be a critical concentration in doped semiconductors was drawn by Mott on the basis of his earlier work [13], where he suggested for the first time that such a concentration should exist for a lattice composed of one-electron (or monovalent) atoms.

According to the one-electron band theory, which can, within certain limits, describe quite well the properties of the majority of crystals, the lattice of monovalent atoms should be metallic for any parameter, even when the overlap of the wave functions of neighboring atoms is very slight. The number of states in the conduction band is twice the number of electrons (because of the two possible directions of the electron spin) and half the states remain free, ensuring metallic conduction. According to Mott, such a lattice is a metal only if the distance between atoms d is less than a certain critical value d_0. If $d > d_0$ or, in other words, at concentrations below the critical value, a crystal should have insulating properties. The transition from an insulator to a metal on reduction of the distance between the atoms is attributed by Mott to a sharp increase in the number of free electrons (we shall confine our attention to n-type semiconductors) from zero to a critical value n_{cr}, governed by a numerical criterion deduced by Mott [12, 14]:

$$n_{cr}^{1/3} a_0 \approx 0.25, \tag{1.1}$$

where $a_0 = \hbar^2 \varkappa / m^* e^2$; $\varkappa$ is the permittivity; m^* is the effective mass of an electron.

Mott derived this criterion by considering the following two limiting cases. If there are no free electrons, the binding of a delocalized (detached from an atom) electron with a positive hole remaining at the atom is described by the potential $V_1 = -e^2/\varkappa r$. In the Coulomb interaction of this kind, an electron and a hole always form a bound pair in the lowest-energy state and, therefore, they cannot participate in conduction at $T = 0$ [12, 14].

If the number of free electrons is sufficiently high, the potential V_1 can be replaced with the screened potential V_2:

$$V_2(r) = -\frac{e^2}{\varkappa r} \exp(-qr).$$

According to Mott, there are no bound states if $qa_0 \geq 1$.

The screening constant q is found from the condition [12]

$$q^2 = \frac{4m^* e^2 (3n/\pi)^{1/3}}{\hbar^2 \varkappa}.$$

Hence, we obtain the criterion (1.1) for the critical concentration n_{cr}.

Mott assumed that an insulator–metal transition at the critical electron density should be abrupt because a small number of free electrons cannot exist due to the fact that electrons and holes readily combine to form pairs [2]. The screening mechanism operates only when the critical concentration is reached and this occurs suddenly when the electron density reaches the required value.

These Mott conclusions differ from the band theory because that theory does not allow for the Coulomb interaction between carriers. Although Mott does not give a rigorous mathematical analysis, he himself is of the opinion that this is the cause of the difference between the properties of the lattice described above and those described by the band theory. Mott considers the electron correlation as an attraction between an electron and a localized hole.

Consistent allowance for the Coulomb interaction between carriers was first made by Hubbard on the basis of his approximate model which allowed for the repulsion of electrons located at the same center. This model is currently used widely to describe the effects associated with the existence of the electron correlation [15].

§ 2. Hubbard Model

Hubbard [16] considered a Hamiltonian for a crystal chain composed of one-electron atoms and used the tight-binding approximation; i.e., he employed wave functions for a free atom.

The Hubbard Hamiltonian consists of two parts: $H = H_1 + H_2$. The first part H_1 describes the motion of an electron between neighboring atoms without allowance for the Coulomb interaction between electrons and the value of H_1 is proportional to the energy-band width $\mathscr{E}$. The second part of the Hamiltonian H_2 allows for the fact that, if electrons are located on the same atom, they are repelled and the energy of the system increases by the repulsion energy U, so that $H_2 \sim U$. Hubbard does not allow for the interaction of electrons between different centers.

If the energy balance is considered in a simplified manner, the localization of an electron can be described as follows. Localization increases the kinetic energy of an electron because it moves within a narrower range. This increase in energy is of the order of $\mathscr{E}$, and the value of $\mathscr{E}$ is governed by the distance between atoms, increasing on their mutual approach. On the other hand, in the case of localization at a neutral atom, the potential energy of an electron descreases by an amount $U = I - E_{aff}$, where I is the ionization energy of a neutral atom and E_{aff} is the affinity energy, i.e., the binding energy of an excess electron to a neutral atom. The value of U is equal to the energy of repulsion between two electrons located at the same atom and is independent of the distance between the atoms so that, when this distance is varied, we can have a variety of relationships between U and $\mathscr{E}$.

If $U > \mathscr{E}$, the electron states are localized and a crystal is an insulator. However, if $U < \mathscr{E}$, the material is a metal. The critical value of the ratio $\mathscr{E}/U = 1.15$, deduced from this model, gives the criterion for the insulator–metal transition. It is close to the Mott criterion (1.1) if we assume that the potential energy of the binding between an electron and a hole is $e^2/\varkappa r$ and the kinetic energy of an electron is $\hbar^2/2m^*r^2$.

The Mott and Hubbard descriptions of the phenomena due to the electron correlation are basically similar and can explain the properties of some insulators [17].

The insulator–metal transition which occurs as a result of a reduction of the distance between atoms in accordance with the Hubbard scheme can be regarded as the filling of the energy gap between two subbands, one of which is filled completely and the other empty (Fig. 1). The maximum value of this gap is $I - E_{aff}$ and, when the distance between the atoms is reduced, the gap decreases to zero.

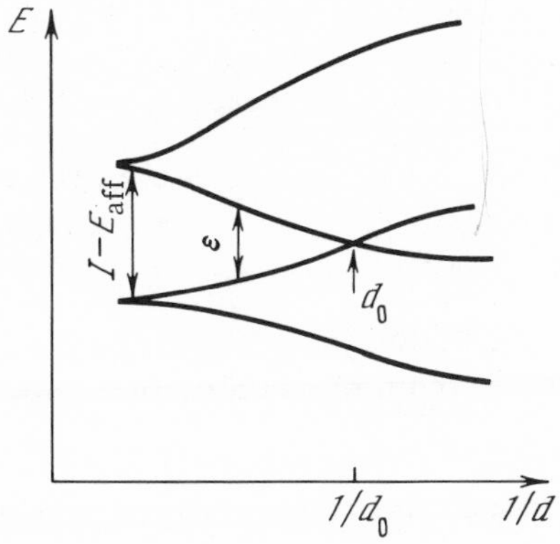

Fig. 1. Overlap of the Hubbard subbands as a result of reduction in the distance between atoms. The distance d_0 corresponds to the Mott transition and $\mathscr{E}$ is the activation energy.

§ 3. Mott Transition in Semiconductors

A metal−insulator transition due to the electron interaction is difficult to observe in its pure form simply by altering the parameters of a crystal under pressure because this approach may result in partial or total destruction of the lattice [18]. However, as mentioned earler, this transition can be observed in doped semiconductors when the average distance between impurities changes as a result of an increase in the impurity concentration.

The Mott transition has been observed in doped semiconductors by many authors and this work was reviewed [1] in a paper presented at the International Conference on Metal−Insulator Transitions held in San Francisco in 1968.

This review gives the results of numerous measurements of the electrical conductivity, Hall coefficient, Hall mobility, magnetoresistance, magnetic susceptibility, nuclear magnetic resonance, and electron spin resonance. The results are compared with the Mott theory. On the basis of these experimental results, the authors of the review put forward a hypothesis of the existence of two critical free-electron densities n_c and n_{cb}.

1. When the electron density is higher than a certain critical value n_c, the electrons become delocalized. If $n < n_c$, the electrons are bound to individual donors. This density n_c agrees with the critical value given by the Mott criterion (1.1).

2. Above a different characteristic density $n_{cb} > n_c$, the Fermi level enters the conduction band.

3. The range $n_{cb} > n > n_c$ is intermediate.

The review gives the results of various measurements of the density n_c in Si and Ge doped with P, As, and Sb. For silicon, this density is $n_c \sim 10^{18}$ cm^{-3} and, for germanium, it is $n_c \sim 10^{17}$ cm^{-3}. For all the experimentally found values of n_c, the product $n_c^{1/3} a_0$ is close to the Mott value ~ 0.25. All the investigations reviewed were carried out on uncompensated semiconductors and, therefore, the densities of free electrons were equal to the donor concentrations.

According to many authors (see, for example, [19, 20]), the merging of the donor impurity band with the conduction band occurs in n-type GaAs when the donor concentration is in the range $3.5 \cdot 10^{16} \leq N_d \leq 2 \cdot 10^{17}$ cm^{-3}.

The dependences of the low-temperature electrical resistivity and of the Hall coefficient on the electron density reveal readily the critical value of this density. When the density is reduced below this value, all these quantities rise steeply [1].

The temperature dependences of the conductivity of doped semiconductors usually consist of several regions [8, 9, 18, 21]: $\sigma = \sigma_1 \exp(-\varepsilon_1/kT) + \sigma_2 \exp(-\varepsilon_2/kT) + \sigma_3 \exp(-\varepsilon_3/kT)$, where $\varepsilon_1 > \varepsilon_2 > \varepsilon_3$ and $\sigma_1 \gg \sigma_2 \gg \sigma_3$.

The activation energy ε_1 is exhibited by samples with electron densities in the range $n < n_c$ and it represents the thermal transfer of electrons from impurity centers to the conduction band.

The activation energy ε_2 is characteristic of densities near the critical value $n \approx n_c$. The energy $2\varepsilon_2$ is that which is required to transfer an electron from one center to another neutral center.

The term with the energy ε_3 occurs only in the conductivity of compensated semiconductors, which we shall discuss in detail later.

§ 4. Anderson Localization of Electrons in Compensated Semiconductors

We shall assume that the degree of doping with shallow donors is so high that a semiconductor is in a state close to degeneracy.

We shall now consider the influence of a second – compensating – impurity introduced into the same semiconductor.

A compensating impurity gives rise to a strongly fluctuating potential in the crystal lattice (see, for example, [22]). The random nature of this potential is much more pronounced than that due to the inhomogeneity of the main impurity distribution.

Materials in which the potential fluctuates strongly are called disordered. Investigations of these materials in the last five years have become an important part of solid state physics. The situation in this branch of physics now resembles that in the physics of pure crystalline semiconductors some 15-20 years ago [3].

The interest in discordered systems arises from practical applications (for example, the applications of the switching effect in amorphous substances) and from the need to extend our theoretical knowledge of the properties of solid and liquid materials with different degrees of lattice disorder.

A characteristic feature of these materials is the absence of the long-range order in the distribution of the component particles so that these materials include a wide range of substances such as liquids and glasses, heavily doped semiconductors, compensated semiconductors, alloys, etc.

Ordered materials, i.e., crystals, have certain universal features of the electron energy structure. Disordered materials should also have such universal features. One is the existence of characteristics of the density-of-states distribution, known as the localized-state "tails" [3, 23].

In considering the question as to whether an electron in a given state is free or localized, we have to deal with two opposite tendencies: electrons can tunnel between the nearest atoms or groups of atoms and they also become localized because of the absence of exact periodicity. The scatter of the energy levels due to this random effect impedes the tunneling.

In 1958, Anderson gave the first quantitative relationship for estimating these two tendencies [24]. The electron levels in an ideal lattice are in regularly distributed potential wells of the same depth which spread out to form a band of width $\mathscr{E}$, whereas, in a random lattice, the wells are different, the levels are spread over a range V_0, and the density-of-states spec-

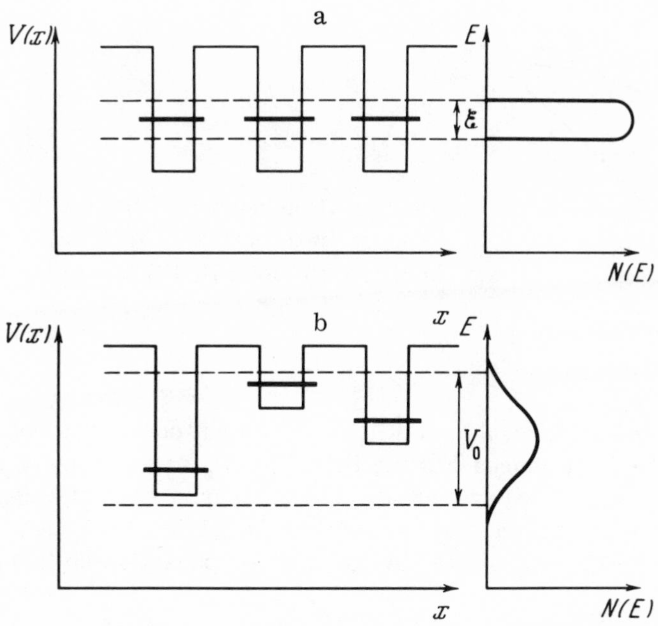

Fig. 2. Potential energy of an electron V(x) and the density of electron states N(E) in the Anderson model: a) for $V_0 = 0$ (periodic lattice); b) for large ratio $V_0/\mathscr{E}$.

trum varies as shown in Fig. 2 [25, 26]. At some critical value of the ratio $(V_0/\mathscr{E})_{cr}$, the states become localized.

There are very many papers dealing with the Anderson localization, i.e., with the localization due to the random nature of the potential (see, for example, [27, 28]). The value of the ratio $(V_0/\mathscr{E})_{cr}$ is a matter of controversy and is currently estimated to be 2-6.

Anderson considered energies in the middle of a band and, therefore, he concluded that, in the case of a strongly random distribution of these energies, the following should apply: 1) all the states in a band should be localized (Fig. 3a); 2) there should be no electron diffusion; i.e., in the limit $T \rightarrow 0$, the mobility should be $\mu \rightarrow 0$ for all the energies inside the band.

Mott [4] combined the arguments relating to the existence of the energy-band tails in disordered materials with the Anderson theory and proposed a band model [3] shown in Fig. 3b. He concluded that, if the Anderson condition was not satisfied throughout a band, it could still be satisfied in the band tails. Therefore, inside the band, there should be a boundary E_c separating the localized and nonlocalized states.

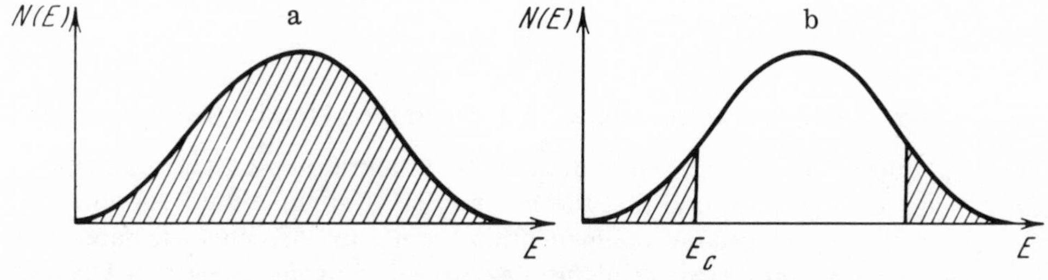

Fig. 3. Anderson band for two different values of the ratio $V_0/\mathscr{E}$: a) greater than critical; b) less than critical. The shaded regions represent localized states.

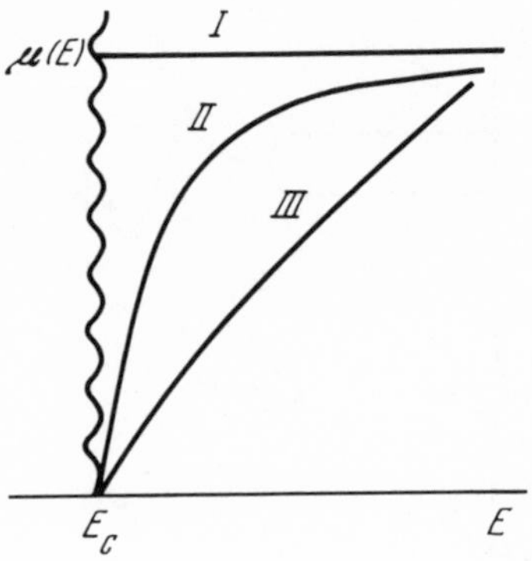

Fig. 4. Dependence of the mobility μ on the energy in the range $E > E_c$ at $T = 0$: I) according to Mott; II, III) according to Cohen.

In the localized-states range, the mobility is $\mu = 0$ but one has to consider the behavior of the mobility when the energy approaches the limit $E \rightarrow E_c$ in the range $E > E_c$. According to Mott [23], this can be studied experimentally in an impurity band of a doped semiconductor if the impurity concentration is such that the semiconductor is on the metallic side of the Mott transition. The Fermi level can be shifted by varying compensation so that this level intersects the boundary E_c.

There are two points of view on the mobility which are illustrated by curves I, II, and III in Fig. 4 [3].

Mott [23, 25, 29] assumes that the mobility should be discontinuous at the point E_c when $T = 0$ (curve I). Cohen et al. [3, 27, 30] describe the behavior of μ (E) as a gradual reduction in the mobility with decreasing energy (curves II and III) on the basis of the percolation theory. When an electron of energy E moves in a random potential $V(\mathbf{r})$, it has allowed regions (A) where $V(\mathbf{r}) < E$ and forbidden regions (F) with localized states where $V(\mathbf{r}) > E$, as shown in Fig. 5 taken from [3]. There is a minimum percolation energy which an electron must have to pass through the whole sample and remain always in classically allowed regions. An increase in the electron energy increases the proportion of allowed regions and, at $E \gtrsim E_c$, they merge into channels penetrating the whole crystal. If $E \gg E_c$, the forbidden regions occupy only a small part in space (Fig. 5). According to these classical ideas, the mobility is almost proportional to the percolation probability, which rises linearly with energy in the range $E \geq E_c$, in agreement with numerical calculations. This corresponds to curve III in Fig. 4. Other calculations give curve II but, in this case, there is no discontinuity of the mobility μ at the point E_c [3].

However, if $\mu(E)$ is described by curve I in Fig. 4, i.e., if we follow the Mott theory, we find that, in the limit $E \rightarrow E_c$, there should be a "residual metallic conductivity," i.e., a conductivity which tends to a finite value in the limit $T \rightarrow 0$. The conductivity is estimated by Mott to be

$$\sigma \approx 0.025/\hbar a \qquad (1.2)$$

or $\sigma \approx 600/a\ \Omega^{-1} \cdot \mathrm{cm}^{-1}$, where a is the distance between the potential wells in angstroms [29].

Mott [25, 29] quotes the experimental data of Davis and Compton [31] in support of his point of view. Davis and Compton find that the resistivity of germanium with a donor concentration of about 10^{18} cm^{-3} and a degree of compensation from 8 to 26% ceases to depend on temperature at ~ 1.5°K, whereas in samples with higher degrees of compensation (up to 80%), there is no saturation of the resistivity at the temperatures employed. The minimum value of the conductivity in the region where it is independent of temperature is close to the value predicted by Eq. (1.2). However, if we assume that the minimum metallic conductivity exists also in samples

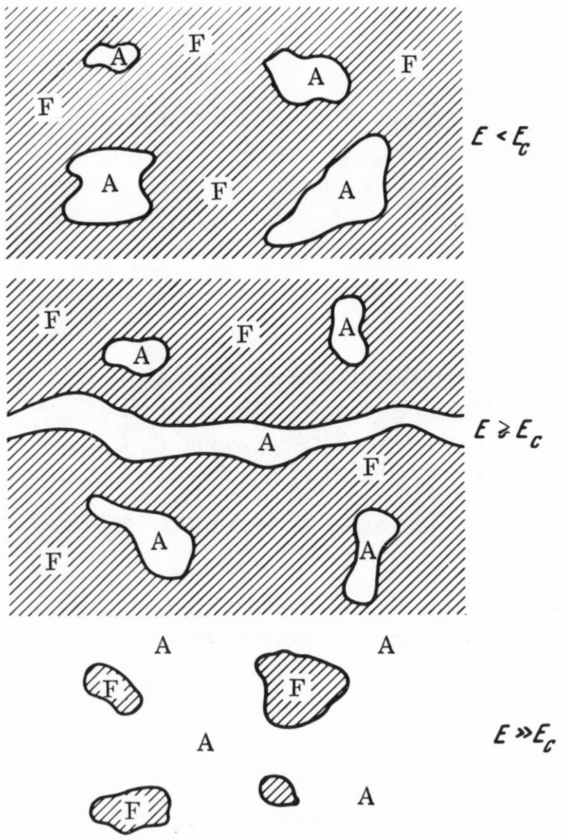

Fig. 5. Allowed (A) and forbidden (F) regions for an electron moving classically in a random potential V(**r**), shown for three different electron energies E. In the allowed regions, we have V(**r**) < E and, in the forbidden regions, we have V(**r**) > E.

which are strongly compensated, and this is found at lower temperatures, extrapolation of the experimental data of Davis and Compton gives a value much lower than that predicted by Eq. (1.2). In fact, the absolute value of the residual metallic conductivity has not yet been determined sufficiently reliably. The value given by Eq. (1.2) is half that suggested by Mott in his earlier papers (for example, [26]) and Mott has tended to assume that it can be less than that given by Eq. (1.2). In one of his recent papers [29], Mott gives the expression $\sigma = \text{const} \cdot e^2/\hbar a_E$ without specifying the constant (here, a_E is the distance between localized states).

The existence of an activation energy at temperatures $T \geq 1.5°K$ in the case of strongly compensated samples (up to 0.8), whose conductivity does not become metallic, is attributed by Mott to the Anderson localization [32]. The dependence of the activation energy on the distance between impurity centers, given in [31], shows that the activation energy of strongly compensated samples does not vanish even when the impurity concentration exceeds the value necessary for a Mott transition in uncompensated samples. On the other hand, the activation energy of weakly compensated samples falls steeply with decreasing distance between the impurity centers and tends to zero at some distance corresponding to an insulator–metal transition, which may be explained by the absence of the Anderson localization.

The disappearance of the metallic conductivity on the introduction of a sufficient amount of compensating impurity into a semiconductor is also reported by other authors (see [6] for references).

Éfros and Shklovskii recently used the percolation model in dealing with electrical conduction in heavily and lightly doped compensated semiconductors with random impurity distributions [33-35]. Éfros and Shklovskii explained the absence of metallic conduction by assuming that the electrons in the conduction band could form metallic drops separated from one another

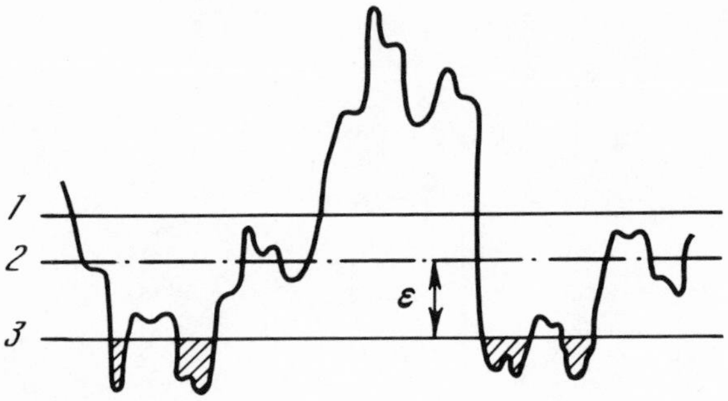

Fig. 6. Energy scheme of a compensated semiconductor: 1) average position of the bottom of the conduction band; 2) percolation level; 3) Fermi level. The shaded regions are occupied by electrons.

by low-permeability barriers; this is shown in Fig. 6, where the shaded regions are similar to the allowed islands in Fig. 5. According to Éfros and Shklovskii [33, 34], the critical degree of compensation at which metallic conduction breaks down is $n_n = N_c^{2/3}/a_0$, where a_0 is the Bohr radius of an impurity and N_d is the donor concentration.

If the compensation is stronger, Éfros and Shklovskii predict activated electrical conduction.

The activation energy, governed by the gap between the Fermi level and the percolation energy at which electrons begin to move across the whole crystal bypassing potential barriers, depends on the compensation of a sample with $n \ll n_c$ as follows:

$$\varepsilon \propto \frac{e^2}{\varkappa} N_d^{1/3} (1 - K)^{-1/3},$$

where $K = N_a/N_d$ is the degree of compensation; N_a is the acceptor concentration.

When the temperature is sufficiently high, the electrical conduction in such a system is due to the thermal transfer of electrons to states with energies exceeding the percolation level. However, at lower temperatures, the suprabarrier motion of electrons becomes difficult and the conduction is due to subbarrier tunneling. The temperature dependences of the electrical resistivity of strongly compensated Ge, CdTe, and InSb are basically identical (see the literature quoted in [6, 29, 35]): the activation energy decreases with decreasing temperature and the resistivity ceases to depend on temperature. This activation energy is regarded by many authors as the energy ε_3 mentioned in §3. According to Mott, the low-temperature dependence of the conductivity is described by [36]

$$\ln \sigma \propto -(\text{const}/T)^{1/4}. \tag{1.3}$$

This dependence has been observed more or less clearly in many investigations ([37-40]; see also references in [6]). According to Shklovskii, at temperatures slightly higher than in the range of validity of Eq. (1.3), when electron jumps are relatively short (see below, § 6), the dependence should be

$$\ln \sigma \propto -(\text{const}/T)^{5/11}, \tag{1.4}$$

which changes to the Mott law (1.3) at lower temperatures when longer jumps are favored by the energy considerations. The Mott law is also valid in the case when there is a correlation between impurities; i.e., this law is more general [40]. The dependences (1.3) and (1.4) have been observed for germanium heavily doped with phosphorus and compensated with gallium [40].

An analysis of the experimental data on compensated semiconductors in the range where the activation energy is ε_3 can be found in [37-41] and this analysis is made on the basis of the hopping mechanism, sometimes even when the conductivity is high (for example, [41]). Weakening of the temperature dependence of the resistivity at lower temperatures, observed in many experiments, is attributed in this analysis to a reduction in the activation energy of the hopping conduction, in accordance with the Mott expression (1.3). Only Mott alone admits the possibility of a different treatment, which attributes this weakening of the temperature dependence to a residual metallic conductivity.

However, the $T^{-1/4}$ law, which according to many authors is the main proof of the existence of the hopping conduction, is always exhibited by compensated semiconductors in a narrow temperature range; since it gives a dependence which is too weak, particularly at low temperatures, it is difficult to regard this proof as final [35]. Some experimenters also report dependences of the $\ln\rho \sim T^{-1/4}$ type but with a proportionality coefficient different from the theoretical value (see [6] for the literature). At low temperatures, the law (1.3) frequently changes to an even weaker temperature dependence of the conductivity [40] and this cannot yet be explained if we assume the hopping conduction mechanism.

Thus, not only a quantitative but even a qualitative interpretation of the experimental data is still lacking.

However, in spite of the lack of clarity in the proposed treatments of the temperature dependence of the conductivity, there is no doubt that compensated semiconductors have an activation energy of this dependence even when the concentrations are sufficient for the Mott transition. This may be explained by the Anderson localization of electrons resulting from the random nature of the potential. The Anderson model is of the one-electron type; i.e., the interaction between electrons is ignored [18].

Another possible cause of the localization of electrons in compensated semiconductors is the Wigner crystallization of electrons resulting from their correlation.

§ 5. Wigner Crystallization of Electrons

In 1938, Wigner investigated the influence of the electron interaction on the energy levels in a metal in the limit of low electron densities and predicted that, in this limit, the electron gas should tend to become localized, forming a regular nonconducting lattice [42]. Wigner used the jellium model, in which electrons were regarded as negative point charges embedded in a uniformly distributed "smeared out" positive charge. Strongly compensated semiconductors with electron densities in the range $n = N_d - N_a \ll N_d, N_a$, moving in a field of randomly distributed positive donors separated by an average distance, which is short compared with the distance between electrons $n^{-1/3}$, approximate closely to this model [2]. In spite of the fact that this has been known for a long time [12] and in spite of the considerable growth of the Wigner lattice theory, the theoretical and experimental aspects of the Wigner theory have only quite recently attracted attention (see, for example, [43, 44]). This may have been due to the success of this theory in explaining other phenomena in which the interaction between electrons plays the dominant part (for example, superconductivity, electron–hole pairing in semimetals with overlapping bands).

In the Wigner lattice description, a convenient measure of the density of a free-electron gas is the dimensionless quantity $r_1 = r_0/a_0$, where a_0 is the Bohr radius and $r_0 = (3/4\pi n)^{1/3}$.

The potential energy of the electron interaction is proportional to $1/r_s$, whereas the kinetic energy is proportional to $1/r_s^2$. If the value of r_s is sufficiently large and the electron density is low, the kinetic energy cannot counteract the tendency for the localization of electrons which are sufficiently far from one another; i.e., it cannot prevent localization of electrons at energy minima. This may result in a periodic structure which is a Wigner electron crystal.

Wigner calculated the correlation energy of an electron crystal, defined as the difference between the energy of the ground state found in the Hartree–Fock approximation, and the value calculated more accurately. The Hartree–Fock approximation, which gives the general band scheme, allows only for the correlation in the distribution of the particles associated with the Pauli principle and it ignores the correlation resulting from the interaction between electrons regarded as charged particles. In the limit of low densities, a rough estimate of the correlation energy [45] gives $E_{corr} = -0.88/r_s$ Ry (1 Ry is equal to $m^*e^4/2\hbar^2\varkappa^2$). The dependence of the correlation energy on the electron gas density, in the limit of low value of this density, was calculated more rigorously by Carr et al. ([46]; see also references in [45]). According to this dependence, the characteristic binding energy of a Wigner crystal in the $r_s > 5$ range is $E_{corr} < 0.05$ Ry [47, 48]. The most detailed investigations of the Wigner lattices were reported by Carr in [46], where he considered the specific heat and magnetic properties of such lattices. In particular, Carr demonstrated that an electron crystal should be antiferromagnetic.

The Wigner crystallization of electrons, like the Mott localization, occurs at the critical density of electrons at which they become delocalized. However, the reported calculated values of the critical density are not in agreement: according to Nozières and Pines [49], the critical value is $r_{s.cr} \approx 20$ whereas, according to Carr [46], this quantity is $r_{s.cr} \approx 3$. A Wigner crystal breaks down on reduction in r_s and on increase in temperature; as in a normal crystal, melting takes place when the amplitude of electron oscillations about their equilibrium positions reaches a certain critical fraction of the distance between electrons (Lindemann melting criterion [45]).

Thus, the external manifestation of the Wigner crystallization is similar to that of the Mott localization of electrons (there is a critical electron density) and of the Anderson localization, which occurs at sufficiently low temperatures. However, in the Wigner crystallization, the electron energies are very low (of the order of a fraction of a millielectron-volt), whereas, in the Anderson localization, the activation energy may amount to a few millielectron-volts [31].

March et al. [50, 51] applied the Wigner lattice idea to semiconductors. They proposed a new treatment of the "freeze-out" of electrons in n-type InSb in a magnetic field, reported by Putley [52]. The activation energy of this material in a magnetic field is higher than the threshold value and cannot be explained by the Yafet–Keyes–Adams (YKA) theory [53] of the compression of wave functions in a magnetic field. The activation energy obtained by Putley (about 10^{-4} eV) is an order of magnitude less than that predicted by the YKA theory. This activation energy was explained by Care and March [51], who calculated the threshold magnetic field in which electrons should form a Wigner crystal in compensated n-type InSb and these calculations gave a magnetic field which agreed with the experimental value.

Care and March [51] considered not only the results of Putley but also measurements of the conductivity of n-type InSb carried out by Somerford [54] at He^3 temperatures, which yielded an activation energy of about $7 \cdot 10^{-5}$ eV. The low values of the activation energy obtained by Putely and Somerford were close to the correlation energy of a Wigner crystal with $r_s \approx 4$ [47].

On the basis of this agreement and the agreement between the calculated threshold magnetic field and the experimental value, Care and March put forward the hypothesis of the existence of a Wigner electron crystal in n-type InSb at low temperatures in strong magnetic fields. They regarded conduction in a Wigner lattice as the diffusion of electrons through vacancies in an electron crystal.

Thus, in considering the conduction in compensated semiconductors, we have to allow for the possibility of collective effects, particularly in those cases when the experimental values of the activation energy are very low. The question as to whether a Wigner lattice with its characteristic properties or some other configuration appears in a crystal and whether it governs the conduction mechanism can only be established by further investigations. A Wigner electron crystal can be regarded only as an idealization of the conditions in real experiments.

§ 6. Some Quantitative Characteristics of Compensated Semiconductors

At the 1962 International Conference on the Physics of Semiconductors held in Exeter, Bonch-Bruevich, Kane, and Keldysh put forward independently the idea of density-of-states tails in heavily doped semiconductors. Keldysh and Kane suggested the following formula for the density of states in a heavily doped semiconductor in which the interaction between electrons and impurities is described by the screened Coulomb potential and the total potential varies quite slowly:

$$\rho(E) = \frac{(2m^*)^{3/2}}{2\pi^2\hbar^2} \int_E^\infty \sqrt{V - E}\,F(V)\,dV, \tag{1.5}$$

where F(V) is the distribution function of the impurity potential; m* is the effective mass; E is the energy measured downward from the bottom of the conduction band.

Therefore, the main problem in the calculation of the density of states is to determine the potential relief created by impurities.

Kane [55] and Keldysh [56] demonstrated that the most suitable approximation for the distribution of the potential is the Gaussian function $F(V) \propto \exp(-V^2/U^2)$, and the density of states in the forbidden band should decrease away from the bottom of the conduction band as $\rho(E) \propto \exp(-E^2/U^2)$, where U is the rms potential of the impurities.

Similar dependences were deduced by Morgan [57] and Lucovsky [58].

In the case of impurities distributed randomly inside a volume $\sim r_{sc}^3$, where r_{sc} is the screening radius, the rms fluctuation in the number of particles is $\Delta N \propto \sqrt{N}$, and the rms potential of such fluctuations is [59]

$$U(r_{sc}) = 2\sqrt{\pi}\frac{e^2}{\varkappa r_{sc}}(N_d r_{sc}^3)^{1/2}, \tag{1.6}$$

where N_d is the donor concentration and $\varkappa$ is the permittivity.

In the electron system and for individual impurity atoms, the condition for the Debye screening is the inequality $e\varphi/kT \ll 1$, where φ is the potential created by all the other impurities at the point where a given impurity atom is located.

The Debye screening radius is

$$r_{sc} = r_D = (kT\varkappa/4\pi ne^2)^{1/2}, \tag{1.7}$$

where n is the density of free electrons.

It follows from Eqs. (1.5) and (1.6) that the Debye screening of fluctuations of the potential occurs when [33]

$$kT \gg \frac{e^2}{\varkappa}\frac{N_d^{2/3}}{n^{1/3}}.$$

The stronger the compensation of a crystal, the more difficult it is to satisfy this condition. In very strongly compensated samples, the Debye screening is impossible because there are very few free electrons. However, according to Shklovskii and Éfros [33, 34], the size and amplitude of the fluctuations are still governed by the electron screening if there is no correlation in the impurity distribution. Electrons screen all the fluctuations except for those for which the rms fluctuation of the number of impurities in a region of radius R is greater than the number of electrons located in this region: $(N_d/R^3)^{1/2} > nR^3$. Thus, regions of radii $R < R_c = N_d^{1/3}/n^{2/3}$ are not screened, i.e., instead of the screening radius, we now have the distance $R_c = N_d^{1/3}/n^{2/3}$ and the amplitude of fluctuations of the potential energy of an electron is $U(R_c) = (e^2/\varkappa)N_d^{2/3}/n^{1/3}$. The electron density is then strongly inhomogeneous and, if N_d and n are large enough, the electrons form metallic drops located at the deepest points in the potential relief [33, 34].

The electron screening radius described above governs the screening in samples with a random distribution of impurities. Samples prepared from the melt represent a different case. At high temperatures, impurities can move easily and, therefore, a Debye–Hückel correlation of positive and negative ions is established in the melt [60] and, at lower temperatures, the impurity distribution can be regarded as the same as that which existed at the freezing temperature T_0. The correlation in such a distribution results in the mutual Debye screening of impurities at distances of the order of [60]

$$r_D = (kT_0\varkappa/8\pi N_d e^2)^{1/2}. \tag{1.8}$$

CHAPTER II

DESCRIPTION OF THE SAMPLES AND MEASUREMENT METHOD

§ 1. Investigated Samples

The present paper reports the results of an investigation of the properties of strongly compensated n-type GaAs. Most of the measurements were carried out on samples cut from the same ingot, which was about 100 mm long and in which total impurity concentration was about $5 \cdot 10^{17}$ cm^{-3}; this concentration was constant to within 20% along the ingot. The room-temperature value of the free-electron density n_0 varied smoothly along the ingot from $2.5 \cdot 10^{15}$ to $5 \cdot 10^{16}$ cm^{-3}, which corresponded to a variation of the degree of compensation from 99 to 82%. The main impurity was tellurium and the compensating impurities (according to the results of a mass-spectroscopic analysis) were oxygen and nickel. A study was also made of samples cut from gallium arsenide ingots which were free of nickel. They were doped with tellurium and compensated with oxygen. The total impurity concentration in these samples was within the range $N_i \approx (3\text{–}6) \cdot 10^{17}$ cm^{-3} and the degree of compensation was K = 10–90%. Measurements indicated that the electrical properties of the investigated samples were independent of the nature of the compensating impurity and were governed only by the degree of doping and compensation.

The electron density was deduced from the Hall effect and the impurity concentration N_i from the electron mobility at liquid nitrogen temperature. At this temperature, electrons were scattered by ionized impurities and N_i was calculated in accordance with [61], where the question of the scattering of electrons in compensated semiconductors was discussed particularly. In the calculation of N_i, it was assumed, following [61], that all the centers were singly charged.

The electrical conductivity and Hall effect were determined for more than 30 samples. Information on some of them is given in Table 1.

TABLE 1

Sample	n_0, cm^{-3}	N_i, cm^{-3}	K, %	Sample	n_0, cm^{-3}	N_i, cm^{-3}	K, %
1	$2.5 \cdot 10^{15}$	$5.5 \cdot 10^{17}$	99	11	$2.1 \cdot 10^{16}$	$4.5 \cdot 10^{17}$	91
2	$5.5 \cdot 10^{15}$	$4.5 \cdot 10^{17}$	97	12	$2.2 \cdot 10^{16}$	$4.7 \cdot 10^{17}$	91
3	$6 \cdot 10^{15}$	$6 \cdot 10^{17}$	98	13	$2.6 \cdot 10^{16}$	$5.3 \cdot 10^{17}$	91
4	$7 \cdot 10^{15}$	$5.5 \cdot 10^{17}$	97	14	$2.6 \cdot 10^{16}$	$4.5 \cdot 10^{17}$	89
5	$8 \cdot 10^{15}$	$6 \cdot 10^{17}$	97	15	$4 \cdot 10^{16}$	$5.2 \cdot 10^{17}$	86
6	$1.2 \cdot 10^{16}$	$4.5 \cdot 10^{17}$	95	16	$4 \cdot 10^{16}$	$4.5 \cdot 10^{17}$	84
7	$1.3 \cdot 10^{16}$	$5 \cdot 10^{17}$	95	17	$4.8 \cdot 10^{16}$	$5 \cdot 10^{17}$	82
8	$1.3 \cdot 10^{16}$	$4.5 \cdot 10^{17}$	94	18	$1.5 \cdot 10^{17}$	$5 \cdot 10^{17}$	54
9	$1.5 \cdot 10^{16}$	$4.6 \cdot 10^{17}$	94	19	$2 \cdot 10^{17}$	$4.5 \cdot 10^{17}$	38
10	$2 \cdot 10^{16}$	$4.8 \cdot 10^{17}$	92	20	$1.5 \cdot 10^{18}$	$2 \cdot 10^{18}$	14

Before the measurements, the samples were etched in a polishing solution (3 parts H_2SO_4 + 1 part H_2O_2 + 1 part H_2O). Contacts of pure tin were formed by alloying in 10^{-5} Torr vacuum at about 500°C. Indium connectors were soldered to tin before the measurements.

Three types of measurement were carried out: the electrical conductivity and Hall effect were measured at low and very low temperatures in weak electric fields; the pulse photoconductivity, excited by ruby laser radiation, was determined; the electrical properties were measured in strong electric fields.

§ 2. Measurements of Electrical Properties in Weak Electric Fields

The electrical conductivity and Hall effect measurements were carried out in the temperature range $0.5 \leqslant T \leqslant 300°K$. This range was divided, in accordance with the method used to establish the required temperature, into two regions.

a) Low temperatures: $1.1 \leqslant T \leqslant 300°K$. Measurements were carried out in the temperature ranges 77-63°K, 20.4-14°K, and 4.2-1.1°K in liquid nitrogen, liquid hydrogen,

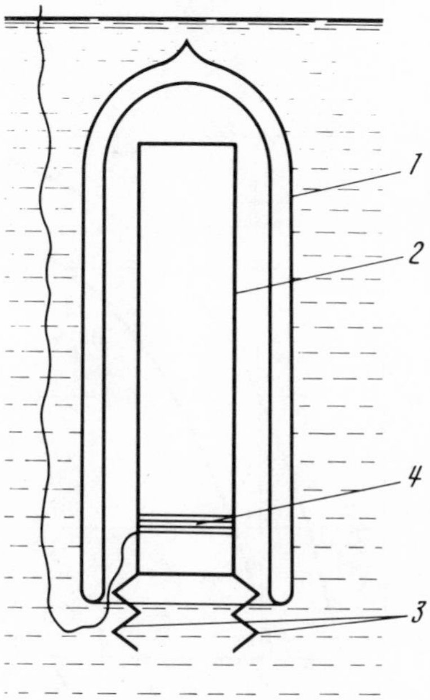

Fig. 7. Inverted Dewar used for measurements in heated helium, hydrogen, and nitrogen vapors: 1) inverted Dewar; 2) copper screen with strips, 3, immersed in cryogenic liquid; 4) heater.

and liquid helium at normal and low pressures of the saturated vapors. In these cases, the temperature of a sample was deduced from the saturated vapor pressure.

Temperatures above the boiling point of these liquids were attained using an inverted Dewar [62], shown in Fig. 7. A tiny inverted Dewar was placed inside a standard liquid helium container and a gas bubble formed inside the Dewar when helium was poured in. A copper foil screen for equalization of the gas temperature was placed inside this Dewar. The bottom of the screen was brought into contact with liquid helium by copper strips. A heater was wound on the lower part of the screen. The screen temperature was governed by the power dissipated in the heater and by the thermal conductivity of the copper strips. A sample and thermometers were placed inside the screen. This arrangement was suggested in [62] for temperatures in the range 4.2-20°K, i.e., only for liquid helium. We applied the same method to liquid hydrogen and liquid nitrogen. When the cryostat was filled with all these liquids in turn, heated vapors could be obtained throughout the range 4.2-300°K. It should be mentioned that it was difficult to heat liquid nitrogen because of the high powers needed for evaporation.

Temperatures in the range $4.2 \leq T \leq 30°K$ inside the inverted Dewar were measured with carbon resistors and those in the range $30 \lesssim T \lesssim 300°K$ were found using automatic copper resistance thermometers.

Temperatures in the range $1.1 \leq T \leq 1.6°K$ were obtained in an apparatus developed at the S. I. Vavilov Institute of Physics Problems of the Academy of Sciences of the USSR: this apparatus is known as the triple Dewar. It differed from a conventional helium cryostat because, inside the usual pair of Dewars (nitrogen and helium), there was an additional helium Dewar, much narrower and with a pump cross-sectional area equal to the cross section of the Dewar itself. The middle helium Dewar acted as the cooling bath for the inner Dewar, which was pumped (by a fast backing pump) to $T \approx 1.1°K$. The small diameter of the inner Dewar reduced the amount of helium carried by the superfluid film from the liquid up along the Dewar and this made the pumping easier [63].

b) Very Low Temperatures: $0.5 \lesssim T \lesssim 1.5°K$. The presence of a superfluid film, which strongly increased the evaporation rate of helium, and the incompatibility of very fast pumps with very narrow Dewars prevented our reaching temperatures lower than $T \approx 1.1°K$ by pumping out He^4, usually employed as the cryogenic liquid. Lower temperatures could be obtained using He^3 as the cooling agent.

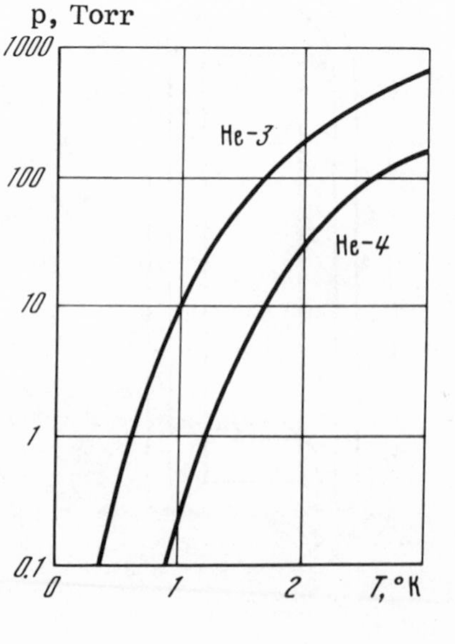

Fig. 8. Temperature dependences of the saturated vapor pressures of He^3 and He^4.

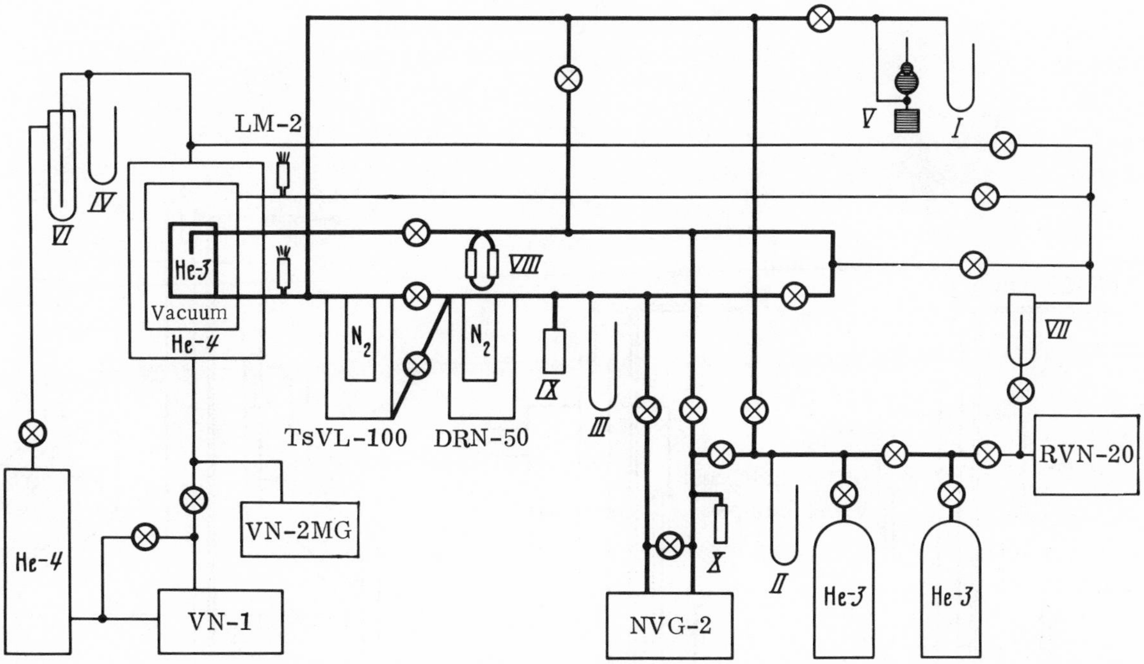

Fig. 9. Gelii-3 apparatus. The part of the system in which He^3 was circulated is shown by thicker lines.

The normal boiling point of He^3 was 3.2°K. Therefore, its vapor pressure at all temperatures was higher than that of He^4 (Fig. 8). When the pressure above the cooling agent was reduced to a certain value, lower temperatures were obtained above He^3 than above He^4. Moreover, He^3 became superfluid only at about 0.008°K [63], so that, down to this temperature, there was no superfluid He^3 film carrying the liquid. These two factors enabled us to achieve very low temperatures by pumping He^3 vapor.

Measurements at very low temperatures were carried out using the Gelii-3 apparatus built on the basis of a design developed in the Cryogenic Division of the Lebedev Physics Institute (a schematic diagram is shown in Fig. 9). This apparatus consisted of three parts.

1. The pumping of He^4 and He^3, purification of He^3 in traps, and its collection in cylinders were all carried out in the operative part. It consisted of two diffusion pumps (TsVL-100 oil pump and DRN-50 mercury pump), four backing pumps (NVG-2 hermetically sealed pump, and RVN-20, VN-2MG, VN-1 pumps), cylinders for gaseous He^3, five traps (VI, VII, VIII carbon traps and IX, X metal traps), and connecting lines with 22 valves.

2. The pressure measurement system comprised four U-shaped manometers (I, II, IV for 760 Torr, and III for 200 Torr), one McLeod gauge (V), and several manometric tubes.

3. The cryostats used in the Gelii-3 apparatus were of three kinds, shown in Fig. 10: A) for measurements in the absence of a magnetic field; B) for measurements using a semiconducting solenoid 7; C) a metal cryostat for measurements in an electromagnet 8. These cryostats were developed in the Cryogenic Division of the Lebedev Physics Institute.

The basic construction of all the cryostats was the same. Gaseous He^3 was cooled and condensed in a closed container 4. Cooling under pumping conditions could result in the stratification of He^3 (below 0.5°K [64, 65]) with the coldest layer of the lowest density and, therefore, located at the top at the liquid–vapor boundary and the warmest layer at the bottom, which

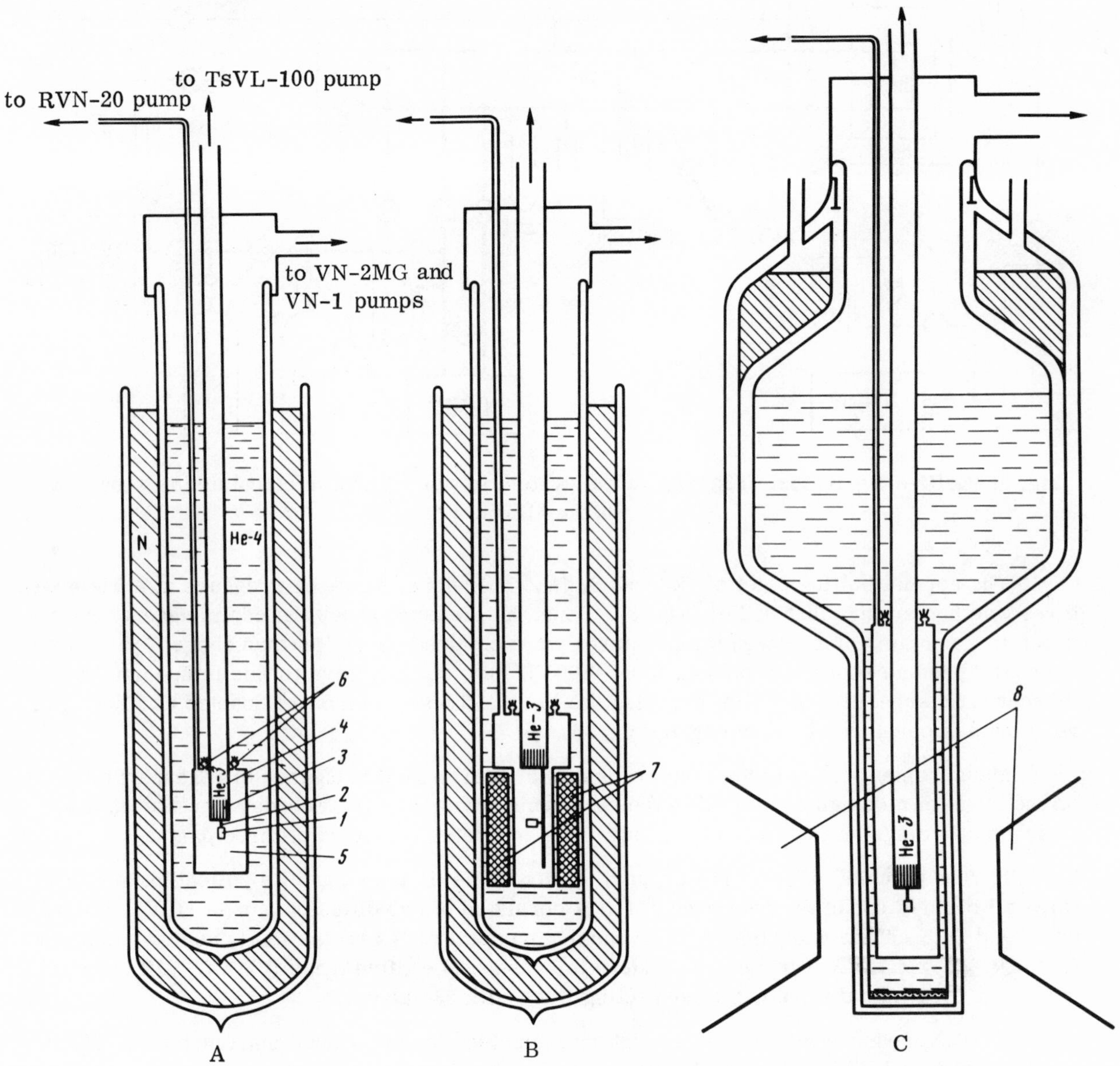

Fig. 10. Cryostats used in the Gelii-3 apparatus

hindered the cooling. Therefore, equalization of the temperature of He^3 was ensured by placing inside the enclosure 4 a condenser 3, which was a comb of very pure copper of high thermal conductivity. A short heat sink was made of the same copper ingot together with the bottom of the enclosure and the condenser; a sample 1 was then bonded to the heat sink. There was no thermal contact between the sample and the liquid He^4 because they were separated by a vacuum jacket 5 in which the residual gas pressure was less than 10^{-3} Torr. Thus, the sample was under the same temperature conditions as the liquid He^3. Therefore, the temperature of the sample could be measured by determining the saturated vapor pressure of He^3 with the McLeod gauge. Moreover, the temperature of the sample was monitored with a carbon thermometer designed for very low temperatures, which was bonded together with the sample to the heat sink.

The vacuum-tight electrical leads 6 were platinum wires sealed into platinum glass connected to the cover of the vacuum jacket by a ferrochrome piece.

When the Gelii-3 apparatus was used, the vacuum jacket, separating thermally the sample from the liquid He^4 bath, was evacuated first with a backing pump. Then, the cryostat was filled in the usual way with He^4 and pumped down to the minimum pressure corresponding to about 1.5°K. The pressure in the vacuum jacket fell to $\sim 10^{-3}$ Torr. Then the He^3 gas was admitted from the cylinders and this gas gradually condensed as a result of cooling, which was deduced with the aid of a mercury manometer. The condensation of He^3 was followed by pumping from the container 4 first by the NVG-2 backing pump and then by the diffusion pumps. The gas returned to the cylinders from the exit of the NVG-2 pump. The admission of He^3 could be repeated, if required, and each time the gas condensation was faster because the reservoir with He^3 was cooled more strongly in each run.

Measurements of the electrical conductivity and Hall effect were made by the usual compensation method at all temperatures.

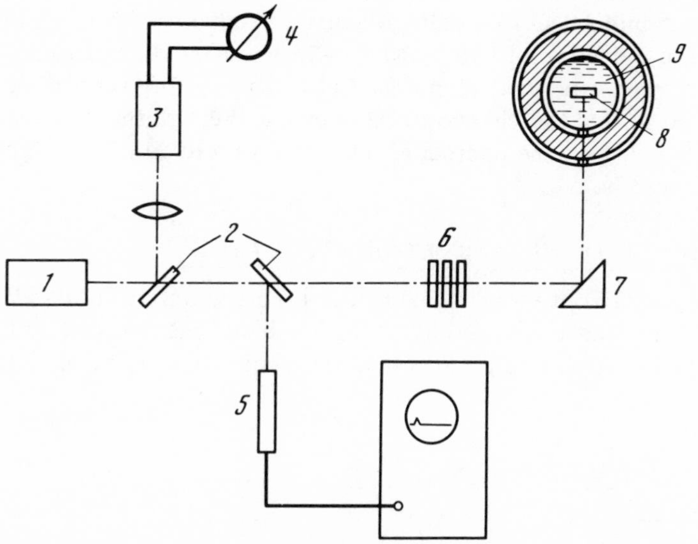

Fig. 11. Illumination system: 1) ruby laser; 2) plane-parallel glass plate; 3) calorimeter; 4) galvanometer; 5) photoresistor; 6) optical filters; 7) rotatable prism; 8) sample; 9) cryostat.

§ 3. Photoconductivity Measurements

We determined the photoconductivity excited by ruby laser radiation using apparatus described in detail in [66]. This apparatus was designed for room-temperature measurements. In the present study, it was used at helium temperatures with the sample illuminated through slits in a glass cryostat, and the electrical connections were provided by a high-frequency cable.

The method used for illumination is illustrated in Fig. 11. A sample was illuminated with single light pulses of $t_p = 4 \cdot 10^{-8}$ sec duration produced by a Q-switched ruby laser. The emission wavelength of the laser was $\lambda = 0.6943\ \mu$ and the photon energy was 1.78 eV.

The intensity of the incident light was measured by deflecting some (10%) of the radiation to a calorimeter with a recording galvanometer using a plane-parallel glass plate inserted in the laser beam at an angle of 45° relative to its axis. An additional plate was used to deflect part of the radiation to a photoresistor so as to obtain a signal for triggering the oscillograph. The intensity of light was varied by altering the voltage applied to the laser and by using a set of calibrated glass filter. In this way, the light flux intensity could be varied from $1 \cdot 10^{18}$ to $5 \cdot 10^{20}$ photons $\cdot$ cm^{-2} $\cdot$ sec^{-1}. A rotatable prism turned the laser beam through 90° and it reached, via the slits in the glass Dewars, the sample inside the helium cryostat.

Reflection from the polished surface of an n-type GaAs sample, at the wavelength employed, was approximately 32% [67] and each glass surface reflected 4% of the incident light. The total reflection from the sample, prism faces, and the walls of the nitrogen and helium Dewars was 55%; reflection from the filter surfaces was allowed for in their calibration.

The cross-sectional area of the laser beam was approximately 1.5 cm^2 on the outer wall of the cryostat. A sample was placed in the central most homogeneous and brightest part of the illuminated spot. The distribution of the illumination intensity in the spot was fairly uniform [66]. The samples were illuminated completely, with the exception of the contacts.

The photocarrier density was found by measuring the photoconductivity using the circuit shown in Fig. 12. A sample was connected in series with a load resistance $R_L \ll R_s$ (R_s was the resistance of the sample) and to a voltage source producing 1.5 V. A signal produced by the load was applied to a system of UR-3 and UZ-5A wide-band amplifiers and then to an S1-11 oscillograph whose screen was photographed. Calibration of the recording system and the value of R_L were used to find the change in the current through the sample as a result of illumination $\Delta I_0 = \Delta V/R_L = I$, i.e., the photocurrent. The amplitude of the signal, recorded photographically, was taken to be ΔV.

§ 4. Measurements in Strong Electric Fields

The electrical conductivity was measured in strong electric fields at 300, 77, 20.4, 4.2, and 1.8°K; the Hall effect was measured at 4.2°K. Heating was avoided by the use of pulse electric fields. The pulse duration was 20 μsec and the repetition frequency was 100-200 sec^{-1}.

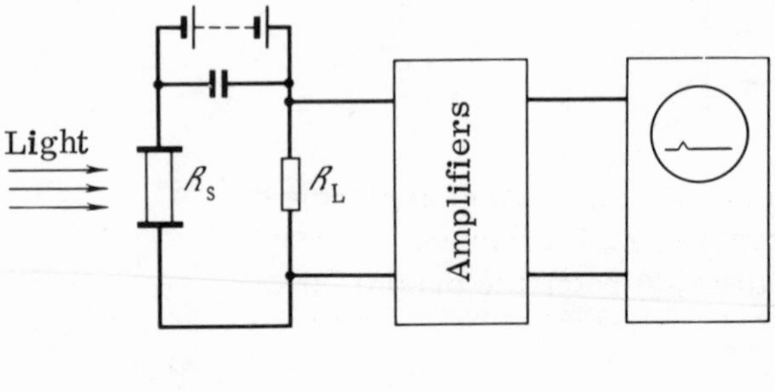

Fig. 12. Photoconductivity measurement circuit.

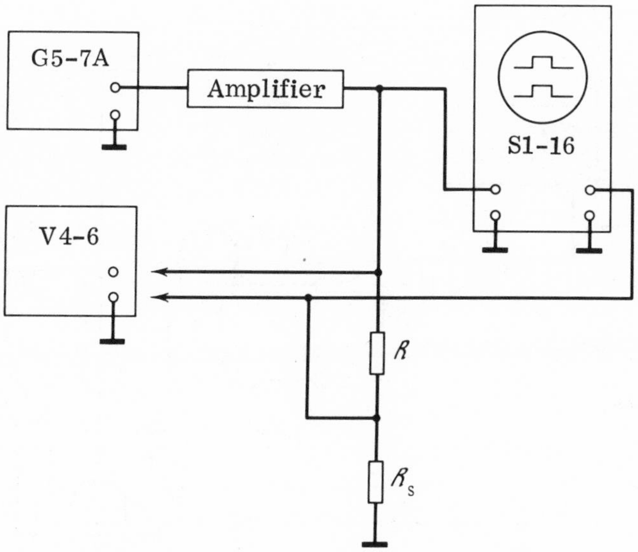

Fig. 13. Circuit for determination of current-voltage characteristics in pulse electric fields.

Measurements were also carried out using other pulse durations and repetition frequencies to check whether heating occurred. Since all these measurements gave approximately the same results, we assumed that the temperature remained practically equal to the temperature of the ambient medium in all cases. This was confirmed by the static current–voltage characteristics of thin samples made of the same material. When the sample thickness was d = 30 μ, there could be no heating because the power dissipated in such a sample would have been too low. The current–voltage characteristics were found to be similar to the characteristics of long samples subjected to electric field pulses.

The pulse current–voltage characteristics were determined using apparatus shown schematically in Fig. 13. The voltage across the sample, which was transmitted via an amplifier from a G5-7A pulse generator, was measured directly with a V4-6 millivoltmeter. The current through the sample was calculated from the voltage drop across a resistor in series with the sample. The shape of the pulses was determined by displaying them on the screen of an S1-16 double-beam oscillograph.

Measurements of the Hall coefficient and resistivity in pulse electric fields were carried out using a specially assembled circuit which was analogous to that shown in Fig. 14 and described in [68]. Pulses of opposite polarity from two symmetric (relative to the ground) outputs of a G5-7A rectangular pulse generator were applied to the current contacts of the sample. In the circuit shown in Fig. 14, the signals from the side contacts were passed to dividers, but in the circuit used in the present study they were applied directly to two identical cathode followers T_1 and T_2 and a subtraction circuit T_3, which subtracted pulses of the same polarity and summed pulses of opposite polarity (the latter were used in the resistivity measurements). A recording instrument V (a V4-6 pulse millivoltmeter or an S1-16 oscillograph) was connected to the output of the circuit. In the absence of a magnetic field, it was possible to equalize, with a potentiometer R, the amplitudes of the pulses of the same polarity, reaching the subtraction circuit from the Hall probes. Thus, during subsequent measurements in a magnetic field, the voltage appearing because of asymmetry in the position of the Hall electrodes was eliminated. The application of a voltage from a calibrated resistor, connected in series with the sample,

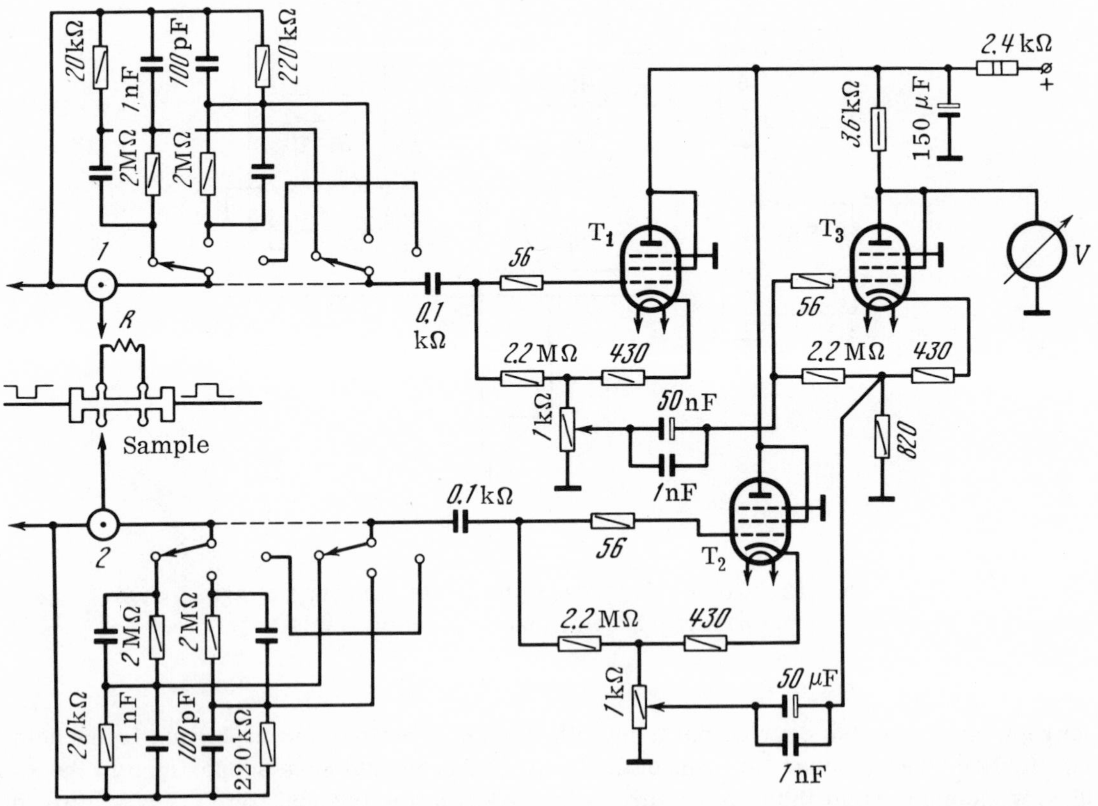

Fig. 14. Circuit for electrical resistivity and Hall coefficient measurements in pulse electric fields.

to the terminals 1 and 2 made it possible to calculate the current in a pulse from the readings of the voltmeter V.

Comparative measurements of the resistivity and Hall coefficient of high-resistivity n-type germanium were made using the circuit described and also under static conditions; this was done to determine the precision of the measurements made using this circuit. The comparison demonstrated that the results were identical to within ±10%.

CHAPTER III

ELECTRON DENSITY AND MOBILITY. CONDUCTION BAND STRUCTURE

We shall report the results of investigations of the resistivity, electron mobility, and electron density in our samples, which were carried out in a wide temperature range $0.5 \leqslant T \leqslant 300°K$ [69, 70]. This range can be divided in a natural manner into three intervals with clearly identifiable characteristics.

§ 1. Temperature Interval $10 \leqslant T \leqslant 300°K$

a) Experimental Results of Hall Effect and Electrical Resistivity Measurements. The results of measurements of the temperature dependence of the resistivity ρ are plotted in Fig. 15. It is clear from this figure that the dependences $\rho = f(T)$

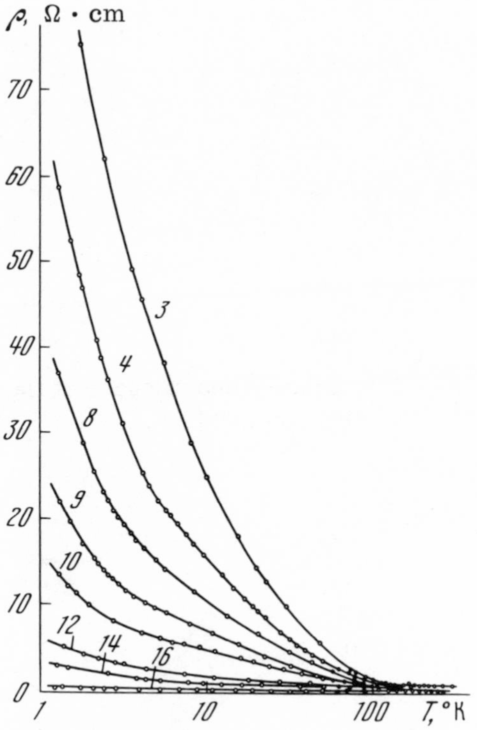

Fig. 15. Temperature dependences of the resistivity. The curves in this figure and all subsequent figures in Chapter III are labeled in the same way as samples in Table 1.

form a family of curves which diverge at low temperatures. The lower the electron density, the higher the rise of the resistivity with decreasing temperature.

The temperature dependence of the Hall coefficient R in the range $8 \leq T \geq 300$°K is plotted in Fig. 16. In this range the samples with electron densities from $6 \cdot 10^{15}$ to $1.5 \cdot 10^{16}$ cm^{-3} exhibit identical $\ln R = f(1/T)$ dependences: near 80°K, the Hall coefficients has a small maximum and then it falls slowly with decreasing temperature, becoming constant at $T \leq 20$°K, where the value of R is approximately equal to the value at room temperature.

Since the change in R is small (not exceeding a factor of 2), the temperature dependence of the resistivity ρ in the range $10 \leq T \leq 300$°K is mainly due to a change in the carrier mo-

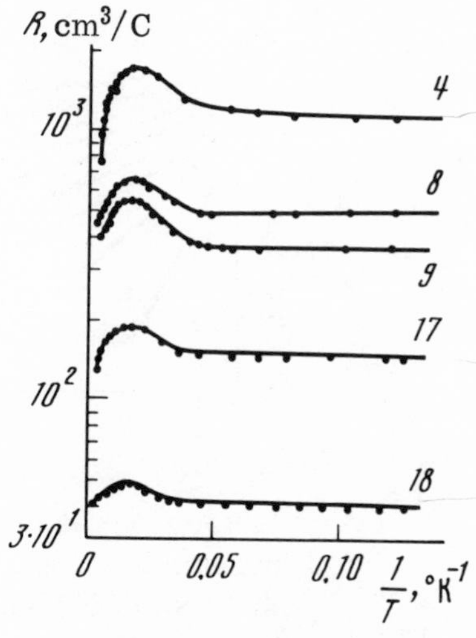

Fig. 16. Temperature dependences of the Hall coefficient.

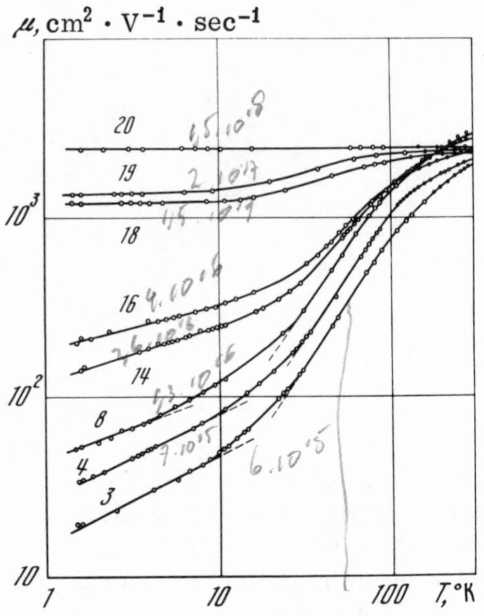

Fig. 17. Temperature dependences of the electron mobility.

bility: in the most strongly compensated samples the mobility changes by a factor of more than 100.

The temperature dependences of the electron mobility μ, calculated from the measurements of ρ and R, are plotted for various samples in Fig. 17. The mobility obeys the law $\mu \propto T^{\delta}$ in the range $20 \leq T \leq 100$°K. In the case of samples with an electron density $n_0 \leq 10^{16}$ cm^{-3} the power exponent in this law is $\delta = 1.5$. The value of δ decreases with increasing electron density, falling to 1.2 for $n_0 = 5 \cdot 10^{16}$ cm^{-3}. For samples with $n_0 > 5 \cdot 10^{16}$ cm^{-3} the dependence $\mu(T)$ becomes weaker with increasing n_0 and when $n_0 = 10^{18}$ cm^{-3} is reached, the mobility remains practically constant throughout the investigated temperature range.

b) Temperature Dependence of the Hall Factor r. The temperature dependence of the Hall coefficient at liquid-nitrogen temperatures can be attributed to the temperature dependence of the Hall factor r in the dependence R = r/ne (for an n-type semiconductor). The Hall factor is given by $r = \langle \tau^2 \rangle / \langle \tau \rangle^2$, where τ is the relaxation time and $\langle\ \rangle$ denotes averaging with respect to the carrier energy [7]. The Hall factor r is related to the power

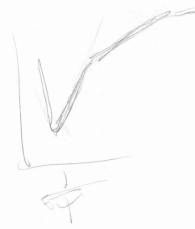

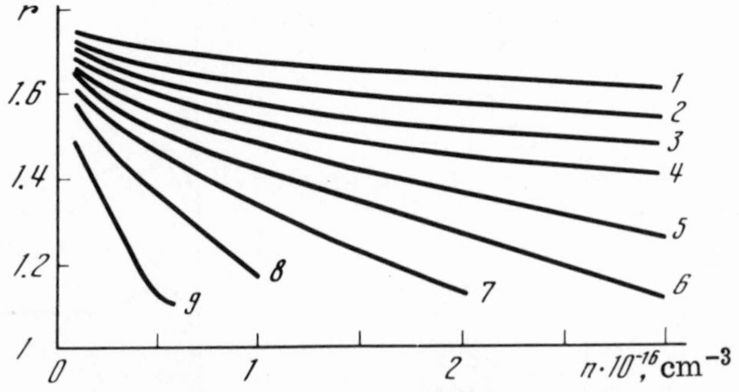

Fig. 18. Temperature dependences of the Hall factor r on the electron density at various temperatures T (°K): 1) 200; 2) 140; 3) 100; 4) 80; 5) 60; 6) 50; 7) 40; 8) 30; 9) 20.

exponent z in the energy dependence of the relaxation time $\tau \propto E^z$. Therefore, r is a function of the electron scattering mechanism and may depend on temperature. At high temperatures the scattering by optical lattice vibrations predominates in gallium arsenide. When this material is cooled, the scattering by optical vibrations in a sample with large numbers of impurities is accompanied by the scattering on ionized impurities. Ehrenreich [71] reports the results of a calculation of the temperature dependence of the mobility for this mixed scattering mechanism. Since even at about 300°K the mobility in our samples does not rise as a result of cooling, in contrast to Ehrenreich's observations, the scattering by impurities is important at all temperatures T < 300°K. Unfortunately, the temperature dependence of r in the range of mixed scattering by the optical vibrations and ionized impurities, 80 ≤ T ≤ 300°K, cannot be determined theoretically since in the case of scattering by the optical vibrations it is not possible to introduce rigorously a relaxation time in this temperature range [71]. We can only calculate the values of r at the extreme points T = 300°K and T = 80°K assuming that at 300°K the scattering by the optical vibrations predominates and the scattering by impurities is dominant at T = 80°K. Then, the Hall factor is r = 1 at T ≈300°K, whereas at T ≤ 80°K, when the scattering by ionized impurities predominates, it follows from [72] that

$$r = \frac{315}{512}\pi\left[\frac{\ln\dfrac{24\pi m^*\varkappa(kT)^2}{ne^2h^2} - 1}{\ln\dfrac{36\pi m^*\varkappa(kT)^2}{ne^2h^2} - 1}\right]^2. \tag{3.1}$$

Here, n is the free-electron density; $\varkappa$ is the permittivity; m^* is the effective mass of an electron.

The factor r was calculated for $m^* = 0.07m_0$ and $\varkappa = 13.5$. The results of the calculations are plotted in Figs. 18 and 19. It is clear from these figures that at T ≈ 80°K the value of r is close to 1.5. Cooling to helium temperatures reduces the r factor to ~1. The dependence $r = f(T)$ calculated from Eq. (3.1) describes well the experimentally observed variation of the Hall coefficient in the range T ≤ 80°K. If allowance is made for the temperature dependence of the Hall factor in the range 10 ≤ T ≤ 80°K, the electron density remains practically constant, equal to n_0 at T = 300°K.

On the other hand, between 80 and 300°K the Hall factor can only vary continuously from 1.5 to 1 if, as usual, the room-temperature scattering in gallium arsenide is dominated by the optical phonons. Therefore, the experimental data on the variation of R by a factor of approximately 1.5 in this temperature range indicates that at 80 < T < 300°K as well as T < 80°K the dependence R(T) is governed by $r = f(T)$. Thus, we may assume that n ≈ const throughout the range 10 ≤ T ≤ 300°K. Table 2 gives the comparative data on 1/ne at T = 4.2°K and at temperatures T_{max} corresponding to the maxima of the dependences $R = f(T)$ deduced allowing for the temperature dependence of the Hall factor r.

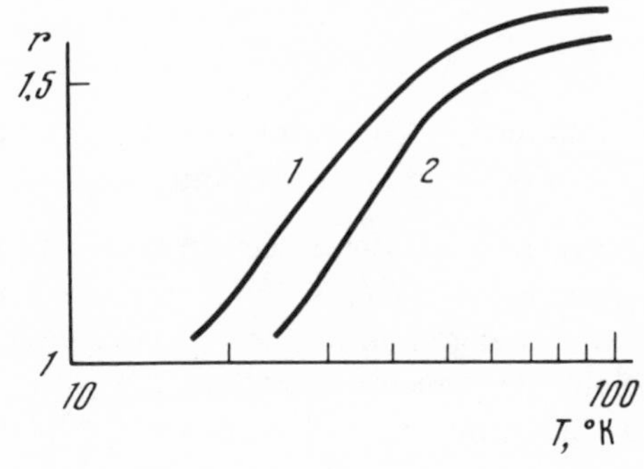

Fig. 19. Temperature dependences of the Hall factor ($m^* = 0.07m_0$): 1) $n = 5 \cdot 10^{15}$ cm^{-3}; 2) $n = 10^{19}$ cm^{-3}.

TABLE 2

Sample	n_0(T = 300°K), cm^{-3}	T_{max}, °K	r	R_{max}	$R_{4,2°K} = \frac{1}{ne}$	$\frac{R_{max}}{r} = \frac{1}{ne}$
4	$7 \cdot 10^{15}$	65	1.53	1750	1150	1145
8	$1.3 \cdot 10^{16}$	70	1.46	675	485	460
9	$1.5 \cdot 10^{16}$	65	1.43	570	400	400
10	$2 \cdot 10^{16}$	65	1.38	425	305	307

It is clear from the reported data that with a reasonable accuracy we can assume that $R_{max}/r = R_{4.2°K} = 1/ne$.

A similar explanation of the dependence $R = f(T)$, obtained under the conditions such that R_{max} exceeds $R_{300°K}$ by a factor slightly less than 2, is given in [73, 74]. On the other hand, some investigators [19, 20, 75] always attribute the maximum of the Hall coefficient of n-type GaAs to the transfer of electrons from the conduction to the impurity band, which is clearly impossible in our case.

c) Temperature Dependence of the Electron Mobility. The temperature dependence of the mobility in the range $20 \leq T \leq 100°K$ generally agrees with the Falicov and Cuevas (F–C) calculations [61] of the scattering by charged centers in compensated semiconductors:

$$\mu_{F-C} = \frac{A(T)}{\ln(1+\eta) + \eta(1+\eta)^{-1}}, \tag{3.2}$$

where

$$A(T) = 2^{7/2}\pi^{-3/2}\varkappa^2(kT)^{3/2}e^{-3}(m^*)^{-1/2}(2N_a)^{-1},$$

$$\eta = \frac{6m^*kT}{\pi^{2/3}\hbar^2(N_d - N_a)^{2/3}},$$

N_d is the donor concentration, N_a is the acceptor concentration, and $2N_a \approx N_i$.

The above formula describes the temperature dependence of the mobility in our samples better than does the Conwell–Weisskopf formula (C–W) [76]:

$$\mu_{C-W} = A(T)/\ln(1+x^2), \quad x = 3\varkappa kT/e^2N_i \tag{3.3}$$

or the Brooks–Herring formula (B-H) [77]:

$$\mu_{B-H} = \frac{A(T)}{\ln(1+y) - y(1+y)^{-1}}; \quad y = \frac{6m^*\varkappa(kT)^2}{\pi\hbar^2e^2n}, \tag{3.4}$$

where n is the free-electron density.

The above three formulas differ in respect of the denominators, which is the result of different allowances for the screening of the Coulomb field of ionized impurities.

Conwell and Weisskopf limited the range of existence of the potential of an impurity center to a radius equal to half the distance between impurities: $r_{0C-W} = (1/2)N_i^{-1/3}$. Brooks and Herring assumed that free carriers could collect around a charged center and screen partly its potential. They found that in the absence of degeneracy the screening radius should be $r_{0B-H} = \varkappa^{1/2}(kT)^{1/2}/2\pi^{1/2}en^{1/2}$. Falicov and Cuevas [61] made the first attempt to allow for the

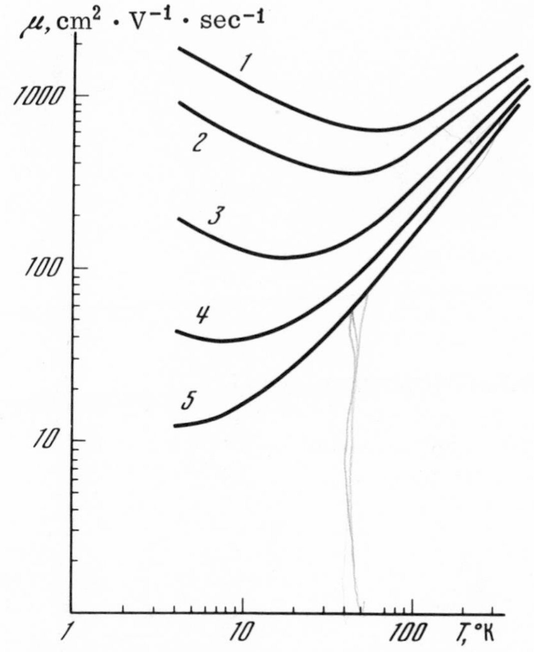

Fig. 20. Temperature dependences of the mobility calculated from Eq. (3.3) for different impurity concentrations (cm^{-3}): 1) $3 \cdot 10^{18}$; 2) 10^{18}; 3) 10^{17}; 4) 10^{16}; 5) 10^{15}.

pair correlation in the distribution of impurities. They obtained a temperature-independent correlation length $a_{F\text{-}C} = 1/r_{0F\text{-}C} = 2\pi^{1/3}(N_d - N_a)^{1/3}$. Therefore, in the limit $T \to 0$ the Falicov–Cuevas formula does not lead to difficulties, in contrast to the Brooks–Herring case when we find that $a_{B\text{-}H} = 1/r_{0B\text{-}H} \to \infty$ in the limit $T \to 0$ [33]. The Falicov–Cuevas formula [61] allows, like the Brooks–Herring formula, for the dependence on the compensation, but no such allowance is made in the Conwell–Weisskopf approximation.

If $m^* = 0.07m_0$ and $\varkappa = 13.5$, the formulas for the temperature dependence of the mobility (3.2)-(3.4) can be represented in the form

$$\mu = \beta_{1,2,3} \frac{2{,}3 \cdot 10^{18}}{N_i} T^{3/2},$$

where

$$\beta_1 = \beta_{C-W} = \frac{1}{\ln(1 + x^2)};$$

$$\beta_2 = \beta_{B-H} = \frac{1}{\ln(1+y) - y(1+y)^{-1}};$$

$$\beta_3 = \beta_{F-C} = \frac{1}{\ln(1+\eta) + \eta(1+\eta)^{-1}}.$$

The results of calculations based on these formulas are plotted in Figs. 20-22. A comparison of these calculations with the experimental results shows that at the impurity concentrations in our samples the formulas (3.3) and (3.4) are invalid already at 50-100°K. The formula (3.2) describes the results satisfactorily down to $T \approx 10$°K.

Thus, in the $10 \le T \le 300$°K range the conduction is due to free electrons whose density remains constant but whose mobility is governed by the scattering on ionized impurities.

§ 2. Temperature Interval 1.5 ≤ T ≤ 10°K

a) Electron Localization. It is clear from Fig. 15 that when the temperature of a sample is lowered, the resistivity rises and the rate of this rise is greater at lower electron densities n_0. This is demonstrated more clearly in Fig. 23 which gives the results of mea-

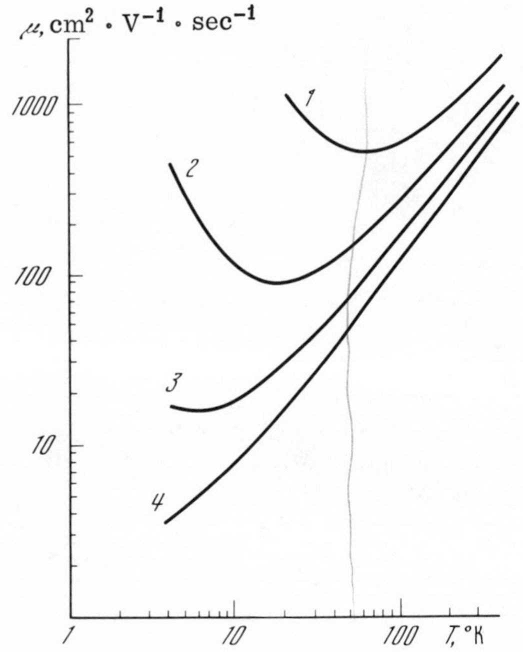

Fig. 21. Temperature dependences of the mobility calculated from Eq. (3.4) for N_i = 10^{18} cm^{-3} and different electron densities (cm^{-3}): 1) 10^{17}; 2) 10^{16}; 3) 10^{15}; 4) 10^{14}.

surements of the resistivity at T = 1.5 and 300°K as a function of the initial electron density n_0 in the samples. It is clear from this figure that at room temperature the electron density and resistivity are related in the usual way by Ohm's law. At T = 1.5°K the resistivity rises much faster with decreasing carrier density. For convenience of comparison of the results of measurements carried out on different samples, we give in Fig. 24 the temperature dependences of the relative resistivities $\rho_T/\rho_{5°K}$. It is clear that the slope of the straight lines in this figure increases with decreasing initial electron density n_0.

The results of measurements of R carried out at T < 5°K are plotted in Fig. 25 in the form of relative values $R_T/R_{5°K}$. The dependences $\ln (R_T/R_{5°K}) = f(1/T)$, like the dependences

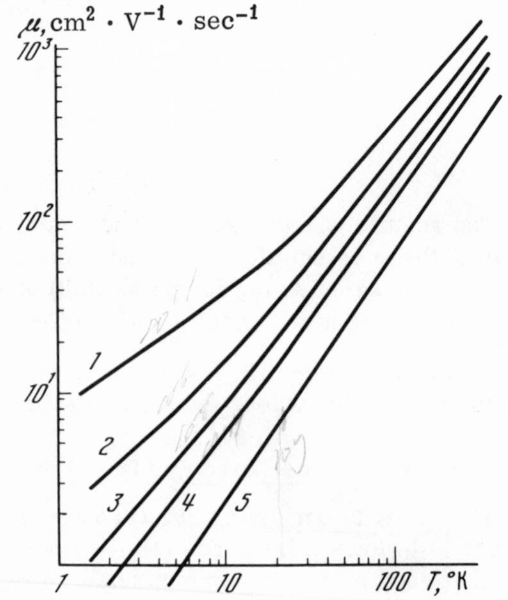

Fig. 22. Temperature dependences of the mobility calculated from Eq. (3.2) for N_i = 10^{18} cm^{-3} and different electron densities (cm^{-3}): 1) 10^{17}; 2) 10^{16}; 3) 10^{15}; 4) 10^{14}; 5) 10^9.

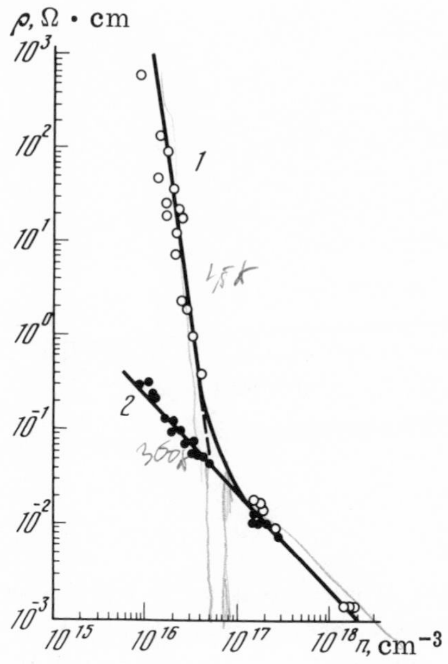

Fig. 23. Dependences of the resistivity on the initial electron density n_0 at T = 1.5°K (1) and at T = 300°K (2).

$\ln(\rho_T/\rho_{5°K}) = f(1/T)$, form a fan of lines whose slopes increase with decreasing initial electron density n_0. The slopes of the lines $\ln(R_T/R_{5°K}) = \varepsilon(1/kT)$ can be used to find the value of ε which can be called the activation energy in this temperature range. The values of ε are very small, of the order of 10^{-4} eV. The lower the electron density at T = 5°K, the higher is the value of ε. The dependence of ε on the electron density n_0 is plotted in Fig. 26. At a certain value of n_0, which is $\sim 5 \cdot 10^{16}$ cm^{-3} in our case, we have $\varepsilon \to 0$. This means that the resistivity and Hall coefficient cease to depend on temperature.

The observed increase in R with decreasing temperature, i.e., the reduction in the free-electron density, may be attributed to the localization of some of the electrons at low temperatures. Such localization may occur at impurity fluctuations, but a rigorous analysis (see below, § 4) meets with difficulties because of the collective effects.

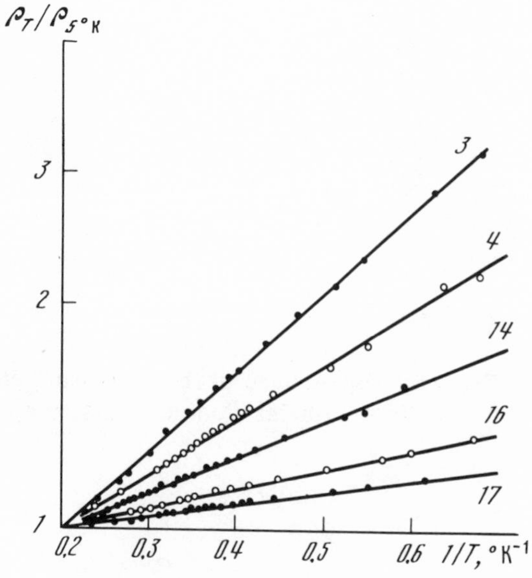

Fig. 24. Relative changes in the resistivity in the range $1.5 \le T \le 5°K$.

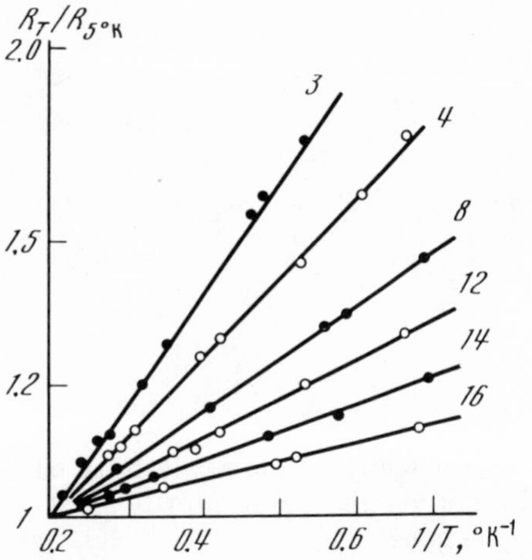

Fig. 25. Relative changes in the Hall coefficient in the range $1.5 \leq T \leq 5°K$.

It is clear from Figs. 23 and 26 that the resistivity and activation energy depend strongly on the electron density. Therefore, we shall assume that a phenomenon analogous to the Mott transition (Chap. I, § 1), in which the screening of the localization centers by free electrons effectively destroys these centers, is observed in our samples at low temperatures. The temperature dependence of the electron density in those samples in which this density is below the critical value may be attributed to the localization of electrons at unscreened localization centers.

b) Scattering by Dipoles. As mentioned in the subsection a) above, the mobility of free electrons is described by Eq. (3.2) down to about 10°K but at T < 10°K the measured values of the mobility are higher than those calculated using Eq. (3.2).

In the case of samples with a relatively low electron density ($n_0 \leq 10^{16}$ cm^{-3}) the experimental dependence is $\mu \propto T^{1/2}$. For a nondegenerate semiconductor, this dependence occurs in the case of scattering by dipoles [78, 79]. Under our conditions at low temperatures, when the de Broglie wavelength of electrons becomes considerably greater than the average distance between oppositely charged centers, the dipoles are increasingly donor–acceptor pairs.

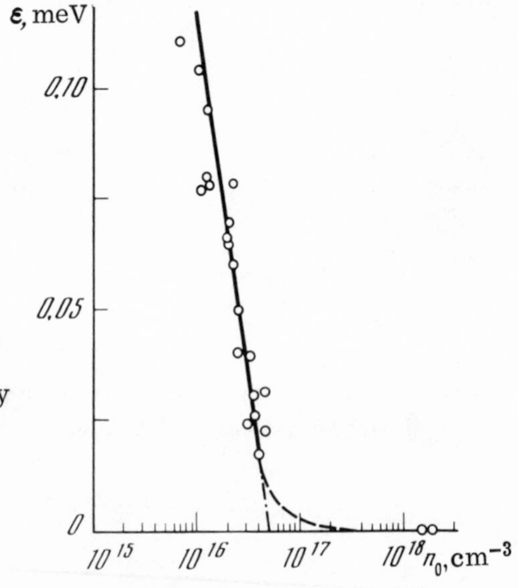

Fig. 26. Dependence of the activation energy ε on the initial electron density n_0.

In our samples the average distance between charged centers is $L_a \approx 1.3 \cdot 10^{-6}$ cm ($L_a^3 N_i = 1$, $N_i = 5 \cdot 10^{17}$ cm^{-3}).

In a semiconductor with sufficiently high and approximately equal donor and acceptor concentrations it is very likely that the neighboring impurity atoms in a cooling melt are oppositely charged. Therefore, we may assume that the dipole length is $L_d = L_a$.

The dipole length L_d can also be taken to be the radius (1.8) of the impurity Debye screening r_D which appears in the melt during the preparation of the samples [60]. If we assume that $T_0 = 1240$°K, $N_d = 2.5 \cdot 10^{17}$ cm^{-3}, and $\varkappa = 13.5$, we find that

$$r_D = \sqrt{\varkappa kT/8\pi N_d e^2} = 1.25 \cdot 10^{-6} \text{ cm}.$$

If we assume that the dipole length is $L_d = 1.3 \cdot 10^{-6}$ cm, we obtain a satisfactory agreement between the measured values of the mobility in the $1.5 \leqslant T \leqslant 10$°K range and those calculated in accordance with the Stratton formula [78]

$$\mu_d = \frac{\sqrt{2}\hbar^2 \sqrt{kT}\varkappa^2}{\pi^{3/2} e^3 (m^*)^{3/2} N_i L_d^2} \frac{1}{f_d(x)}.$$

This calculation was carried out for a sample with the lowest value of n_0 since no allowance was made for the electron screening effect [$f_d(x) = 1$].

Since the dependence $\mu \propto T^{1/2}$ applies only in the absence of degeneracy, we may assume (Fig. 17) that right down to $T \approx 2$°K samples 3-8 remain nondegenerate, samples 18-20 are degenerate, and in samples 14 and 16 the electron density corresponds to a transition between the nondegenerate and degenerate states.

Thus, in the range $1.5 \leqslant T \leqslant 10$°K electrons become localized when the temperature is reduced sufficiently. The remaining free electrons are not scattered by isolated impurities but by impurity dipoles.

§ 3. Results of Measurements at Very Low Temperatures

Measurements of the electrical conductivity and the Hall effect were carried out also at lower temperatures, right down to ~0.5°K, using the Gelii-3 apparatus described in Chap. II,

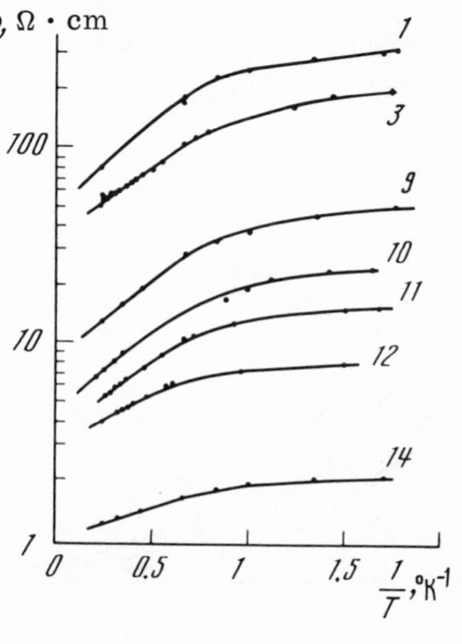

Fig. 27. Temperature dependences of the resistivity at T < 5°K.

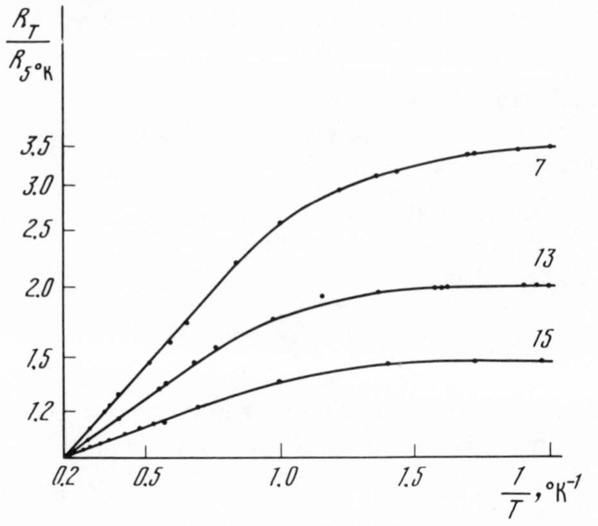

Fig. 28. Temperature dependences of the Hall coefficient at T < 5°K.

§ 2. The results of measurements of the resistivity at temperatures $0.5 \leqslant T \leqslant 5°K$ are plotted in Fig. 27 for a number of samples. We can see that at T < 1°K the resistivity ρ ceases to increase rapidly with decreasing temperature and ρ tends to a limit at $T \rightarrow 0$.

Figure 28 gives the temperature dependences of the Hall coefficient in the form $R_T/R_{5°K} = f(1/T)$ and these are of the same nature as the temperature dependences of the resistivity.

We have mentioned earlier that a reduction in the electron density in the conduction band may be attributed to the localization of some of the electrons and we shall assume that $n(T) = n_0 - n_l(T)$, where n_0 is the total electron density in the conduction band measured at a high temperature; n(T) is the density of electrons in the same band at a temperature T; $n_l(T)$ is the density of localized electrons at a temperature T.

The results of measurements of the localized electron density in the form of the dependence $n_l(T)/n_0 = f(T)$ are represented by points in Fig. 29. Since the $n_l(T)/n_0 = f(T)$ curves in Fig. 29 have a tendency to saturation in the limit $T \rightarrow 0$, the value of $n_l(0)$ can be determined quite accurately by extrapolation of the curves to T = 0. The values of n_0 and $n_l(0)$ are listed in Table 3.

The electron mobility right down to about 1.5-1°K obeys $\mu \propto T^{1/2}$ and at lower temperatures it also exhibits a tendency to saturation, as illustrated in Fig. 30.

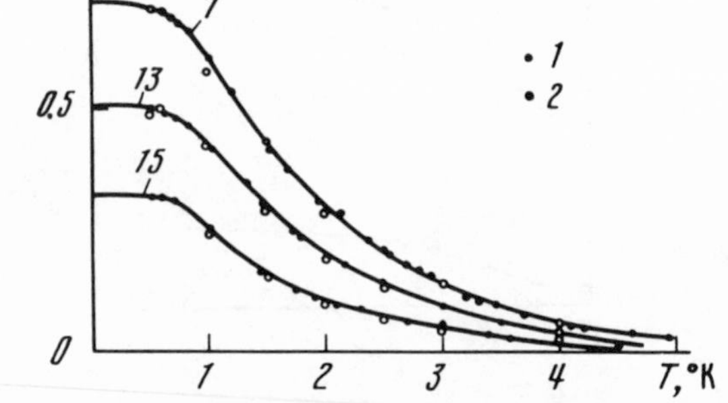

Fig. 29. Temperature dependences of the density of localized electrons, reduced to the initial density n_0: 1) experimental result; 2) calculations.

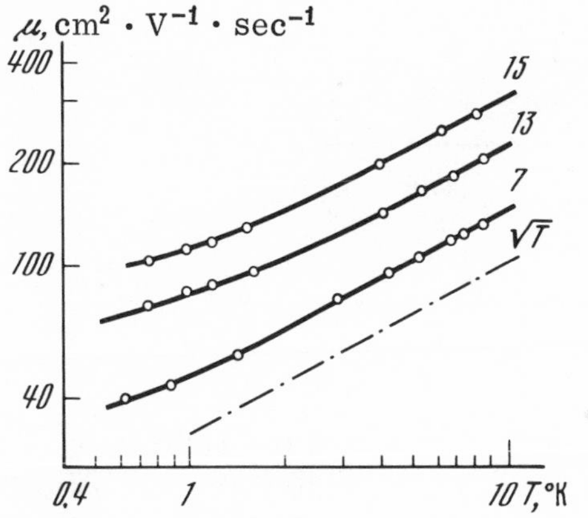

Fig. 30. Temperature dependences of the electron mobility in the dipole scattering range.

§ 4. Physical Mechanism of Electrical Conduction and Some Qualitative Estimates

Difficulties of two kinds are encountered in establishing the physical mechanisms of the phenomena observed in our samples. First of all, we are dealing with the moderate doping case ($N_d a_0^3 \sim 1$), for which the theory is insufficiently developed because of the absence of a small parameter. Secondly, difficulties arise in connection with the correlation in the distribution of impurities. The assumption of a potential relief formed by impurity fluctuations is an important element of the physical picture but it is usually not clear how random is the impurity distribution. This leads to an indeterminacy of some important parameters, like the characteristic size of fluctuations and the closely related energy scale of fluctuations.

Naturally, the motion of an electron is governed, on the one hand, by the potential relief and, on the other, by the interaction of electrons with one another. In the first approximation we shall ignore the collective effects and consider the motion of individual electrons in a field created by impurities.

We shall assume that the impurity distribution is correlated because of the mutual Debye screening of impurities at the melt freezing temperature (Chap. I, § 6). Then, the characteristic fluctuation radius R_{fl} at this temperature is the Debye radius r_D of Eq. (1.8):

$$R_{fl} = r_D = \sqrt{\varkappa k T / 8\pi N_d e^2} = 1.2 \cdot 10^{-6} \text{ cm}.$$

We shall assume that the correlation in the distribution of impurities with the Coulomb potential is equivalent to a random distribution of impurities with a potential screened in the Debye manner. Therefore, the average amplitude of fluctuations of the potential is proportional to the square root of the fluctuation radius (1.6):

$$U_{fl} = \frac{e^2 (4\pi N_d r_D^3)^{1/2}}{\varkappa r_D} \approx 20 \text{ meV}. \tag{3.5}$$

We shall regard a positive fluctuation of the potential as spherical in respect of its spatial coordinates and as a rectangular potential well with a radius r_D and depth U_{fl}. The minimum depth of this well, at which an electron level appears inside it, is [80]

$$U_{min} = \pi^2 \hbar^2 / 8 m^* r_D^2 \approx 8.5 \text{ meV} \tag{3.6}$$

for $m^* = 0.07 m_0$.

TABLE 3

Sample	$n_0 \cdot 10^{-16}$, cm^{-3}	$n_l(0) \cdot 10^{-16}$, cm^{-3}
7	1.3	1.0
13	2.6	1.3
15	4.0	1.3

N_i

5.10^{17}
3.10^{17}
$5.2 \cdot 10^{17}$

A comparison of Eqs. (3.5) and (3.6) shows that $U_{fl} > U_{min}$ and the well depths in our samples are sufficient for electron capture. A well of depth U_{fl} = 20 meV contains only one level because the next one appears when the depth becomes $9U_{min}$ = 76 meV. The Coulomb repulsion between electrons at this single level in the well ensures that the well contains just one electron. Thus, we may assume that the number of potential wells in which electrons may be localized is equal to the maximum number of localized electrons $n_l(0)$. It is clear from Table 3 that this number is of the order of 10^{16} cm^{-3} and that it is approximately the same for all the samples. This circumstance can be explained by the fact that the impurity concentration and the sample preparation technology are the same in all cases. Consequently, the frozen-in fluctuations in the impurity distribution are also approximately similar.

We shall assume that localized states occupy an energy band of width Δ, and that this band is adjacent to the bottom of the conduction band. The width Δ should be of the order of the activation energy ε discussed in § 2. Since — according to our measurements — even at $T \approx 5°K$ almost all the wells are empty, it follows that $\Delta \approx kT \approx k(5°K) \approx 4.5 \cdot 10^{-4}$ eV and an estimate shows that the average density of states in the localization zone $\rho_1 = n_l(0)/\Delta$ is of the order of 10^{31} cm$^{-3} \cdot$ erg^{-1}. The density of states in the free-electron band can be estimated by dimensional analysis [see Eq. (1.5)] as $\rho_2 \propto (m^*)^{3/2}(U_{fl}/\hbar^3)^{1/2}$. In our case, $\rho_2 \sim 10^{32}$ cm$^{-3} \cdot$ erg^{-1}.

The total electron density in our samples is $n_0 = N - N_a > n_l(0)$, i.e., it is greater than the concentration of potential wells that can hold an electron each. In the limit $T \to 0$ the levels in these wells become filled and the excess electrons remain in the conduction band in the form of a frozen gas. This is confirmed by the experimentally observed weakening of the temperature dependence of the mobility $\mu(T)$, as shown in Fig. 30.

Up to now our discussion has been based on average estimated values of the relevant quantities. Clearly, the dimensions and depths of the wells are not all equal. The levels in the wells can be at various heights or they may be absent from the wells. It is this scatter of conditions that governs the number of localized electrons and the width of the localization band Δ. The relevant characteristics can be determined only on the basis of more detailed information on the potential relief created by impurities. More details can be obtained on the basis of the following considerations.

As suggested earlier, we shall assume that the real distribution of charged impurities can be replaced by a random distribution of equivalent impurities creating a screened potential $U(r) = \pm e \exp(-r/r_D)/\varkappa r$. Clearly, the characteristic size of fluctuations is then governed by the Debye radius r_D and the fluctuations of the potential are governed by fluctuations of the charge inside a volume whose characteristic size is $2r_D$.

We shall divide the whole sample into cubes of edge $2r_D$. The number of such cubes per unit volume is $N_0 = 1/(2r_D)^3 = 6.6 \cdot 10^{16}$ cm^{-3}.

We shall calculate the probability that a fluctuation of charge $\Delta Q = ke$, where $k = 0, \pm 1, \pm 2, ..., e > 0$, occurs in such a cube. We shall assume that the probability w_n that a given volume contains n impurities of one sign, i.e., only donors or only acceptors (these probabilities are equal apart from a small degree of decompensation), is described by the Poisson law

$$w_n = \frac{\bar{n}^n}{n!} \exp(-\bar{n}),$$

where

$$\bar{n} = (2r_D)^3 N_d = 3.9$$

is the average number of donors or acceptors in an elementary cube. The charge fluctuations $\Delta Q = ke$ are a consequence of the situation in which a given volume contains n donors and n + k acceptors. Since the distributions of equivalent impurities are independent, the probability of this situation $w_{n,n+k}$ is equal to the product of the probabilities of the relevant events $w_{n,n+k} = w_n \cdot w_{n+k}$.

Summing over all possible values of n, we find that the probability of occurrence of a charge fluctuation $\Delta Q = ke$ in an elementary cube is

$$W_k = \sum_{n=0}^{\infty} w_n w_{n+k} = \sum_{n=0}^{\infty} e^{-\bar{n}} \frac{(\bar{n})^n}{n!} e^{-\bar{n}} \frac{(\bar{n})^{n+k}}{(n+k)!} = e^{-2\bar{n}} \sum_{n=0}^{\infty} \frac{(2\bar{n}/2)^{2n+k}}{\Gamma(n+1)\,\Gamma(n+k+1)} = e^{-2\bar{n}} I_k(2\bar{n}),$$

where I_k is a Bessel function with an imaginary argument.

We note that $W_k = W_{-k}$, i.e., the probability of occurrence of a potential hump is equal to the probability of occurrence of a potential well.

If $\bar{n} = 3.9$, it follows from the standard formulas [81] that $I_k(2\bar{n})$ has the values

$$I_0(7.8) = 355; \quad I_2(7.8) = 270; \quad I_4(7.8) = 121; \quad I_6(7.8) = 35.$$
$$I_1(7.8) = 331; \quad I_3(7.8) = 193; \quad I_5(7.8) = 69;$$

Moreover, $\sum_{k=7}^{\infty} I_k(7.8) = 24$, and it follows from the normalization condition that

$$\sum_{k=-\infty}^{\infty} e^{-2\bar{n}} I_k(2\bar{n}) = 1.$$

Thus, the probabilities of the occurrence of a charge fluctuation $\Delta Q = ke$ in an elementary cube of size $2r_D$ are as follows:

$$W_0 \approx 0.146; \quad W_2 = W_{-2} \approx 0.110; \quad W_4 = W_{-4} \approx 0.050; \quad W_6 = W_{-6} \approx 0.014.$$
$$W_1 = W_{-1} \approx 0.136; \quad W_3 = W_{-3} \approx 0.079; \quad W_5 = W_{-5} \approx 0.028;$$

The probabilities with higher values of k are

$$\sum_{k=7}^{\infty} W_k = \sum_{k=7}^{\infty} W_{-k} = 0.010.$$

The potential created by a fluctuation $\Delta Q = ke$ will be approximated by a spatially spherical well of radius r_D and depth $U_{\Delta Q} = -e\Delta Q/\varkappa r_D = -e^2 k/\varkappa r_D$.

In this case the scale of fluctuations of the potential is $e^2/\varkappa r_D \approx 8.5$ meV.

Therefore, the depths of the wells are

$$U_1 = 8.5 \text{ meV}; \quad U_3 = 25.5 \text{ meV}; \quad U_5 = 42.5 \text{ meV};$$
$$U_2 = 17 \text{ meV}; \quad U_4 = 34 \text{ meV}; \quad U_6 = 51 \text{ meV}.$$

The potential relief is very complex. However, only some of the wells may localize electrons. Since an electron at an energy level E in a well moves within limits given by the relationship $pr_c = (2m^*E)^{1/2}r_c = \hbar$, it follows that r_c can be regarded as the radius of the state with an energy E. If this radius is greater than r_D, an electron cannot be regarded as localized in a well of radius r_D because it may be located also in neighboring wells.

We shall now consider the possibility of localization of electrons in specific potential wells which are likely to occur in our material. A well of depth $U_1 = 8.5$ meV cannot contain an electron because it is too shallow. A well of depth $U_2 = 17$ meV does contain a level corresponding to a bound state. However, the radius of this state is $r_2 \approx 1.1r_D$ and an electron is not localized at this level. A calculation (see the Appendix at the end of the present chapter) shows that an electron may be localized only in wells beginning from U_3 or deeper in which the bound-state radius is less than r_D. The occurrence of wells deeper than U_6 is unlikely. Thus, the localization occurs in wells from U_3 to U_6 and all of them capture one electron each.

The number of such wells is clearly $n_l(0) = \sum_{k=3}^{\infty} W_k N_0 = 0.181 \cdot 6.6 \cdot 10^{16} = 1.2 \cdot 10^{16}$ cm^{-3}, which is very close to the experimental value of $1.3 \cdot 10^{16}$ cm^{-3}.

It is very likely that the electron localization occurs not in individual wells and that the levels of separate wells are mixed to form a local band. Then, the motion of an electron is limited to narrow channels connecting such linked wells. Electrons in these channels need not participate in the dc conduction because these channels do not penetrate the whole crystal.

We shall confine ourselves to these general representations because our model meets with a basic difficulty, which is a low value of the activation energy ε (or of Δ). In fact, the characteristic fluctuation scale is $\Delta U \sim 8.5$ meV. Consequently, the scatter of the levels in the wells U_3-U_6 which can localize electrons is about 10 meV, and, therefore, Δ should be ~10 meV and not a fraction of a millielectron-volt, as found experimentally (§ 2a).

Until now we have ignored the screening effect of free electrons. Such screening clearly exists and it should reduce the width of the localized-states band because of the general smoothing of the potential relief and disappearance of deep levels. Clearly, an increase in the free-electron density should also reduce the width of the localization band; i.e., the activation energy should decrease because of an increase in the screening effects, which is confirmed by the experimental results. However, it is not possible to estimate quantitatively the influence of the electron screening because this would require allowance for the Coulomb repulsion between electrons and the problem would become of the many-body type.

There are also other possibilities which may explain the resultant situation.

1. If the impurity distribution is strongly correlated, the fluctuation scale decreases and low values of Δ may occur.

2. The value of Δ may also decrease because of ordering inside the electron system as a result of collective effects of the kind represented by the Wigner lattice. In this case we cannot speak of localization centers. Nevertheless, we can speak of localization of electrons in the sense that some of them do not participate in conduction.

§ 5. Simplified Energy Model and Calculation of Its Parameters

Since a complete solution cannot be obtained because of the many-body nature of the problem, it is interesting to consider even a very approximate one-electron model of the spectrum.

On the basis of the considerations put forward above, we shall calculate the energy structure of the conduction band using the model shown schematically in Fig. 31.

An energy band of width Δ near the bottom of the conduction band is occupied by states in which electrons are localized. The density of states ρ_1 in this localized band is

$$\rho_1 \Delta = n_l(0), \tag{3.7}$$

where $n_l(0)$ still denotes the density of localized electrons in the limit $T \to 0$. An energy band of density ρ_1 extends over a region E_0 such that

$$\rho_1 E_0 = n_0, \tag{3.8}$$

where n_0 is the total density of electrons in the conduction band.

At the upper edge of the localized band the density of states changes suddenly from ρ_1 to ρ_2. Thus, near the bottom of the conduction band the density-of-states distribution is described by

$$\rho(E) = \begin{cases} \rho_1, & -E_0 \leqslant E < 0; \\ \rho_2, & E \geqslant 0. \end{cases}$$

This step-like approximation to the real distribution allows us to estimate roughly but rapidly the average characteristics of the distribution and it is convenient for the description of the temperature dependence of the electron density. For ease of calculation, it is also assumed that zero energy corresponds to the step in the density of states.

According to the energy scheme of Fig. 31, the density of localized electrons is

$$n_l(T) = \rho_1 \int_{-E_0}^{-E_0+\Delta} \frac{dE}{1 + e^{\xi(E-F)}} = n_l(0) - \frac{\rho_1}{\xi} \ln \frac{1 + e^{\xi(-E_0+\Delta-F)}}{1 + e^{\xi(-E_0-F)}}, \tag{3.9}$$

where $F = F(T)$ is the Fermi level; $\xi = 1/kT$; k is the Boltzmann constant.

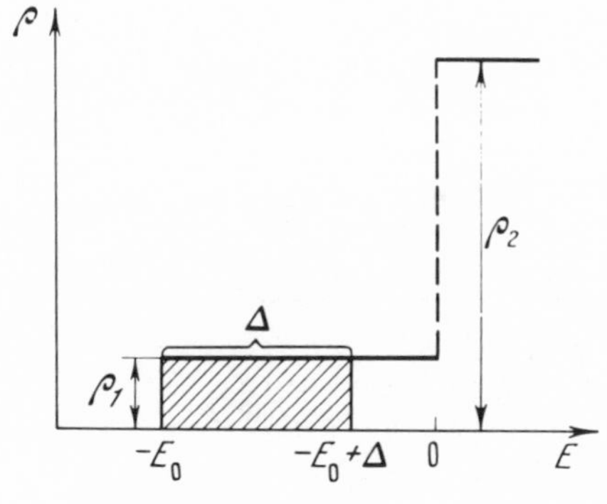

Fig. 31. Energy distribution of the density of states near and in the conduction band.

The total electron density in the conduction band is then

$$n_0 = \rho_1 \int_{-E_0}^{0} \frac{dE}{1 + e^{\xi (E-F)}} + \rho_2 \int_{0}^{\infty} \frac{dE}{1 + e^{\xi (E-F)}}. \tag{3.10}$$

Using Eqs. (3.7) and (3.8), we can transform Eqs. (3.9) and (3.10) to

$$\ln \frac{1 + e^{\xi (-\Delta/f_0 - F + \Delta)}}{1 + e^{\xi (-\Delta/f_0 - F)}} = \Delta\xi \left(1 - \frac{f}{f_0}\right), \tag{3.11}$$

$$\ln \frac{1 + e^{-\xi F}}{1 + e^{\xi (-\Delta/f_0 - F)}} = \frac{\rho_2}{\rho_1} \ln (1 + e^{\xi F}), \tag{3.12}$$

where $f_0 = n_l(0)/n_0$, $f(T) = n_l(T)/n_0$.

This system of equations contains an unknown variable F(T) and two unknowns ρ_2/ρ_1 and Δ, which (according to our hypothesis) are independent throughout the range of temperatures of interest to us. This system can be solved by finding, from Eq. (3.11), the quantity F as a function of T and Δ:

$$(-\xi) F = \ln \frac{e^{\xi\Delta/f_0} (e^{\Delta\xi\psi} - 1)}{e^{\Delta\xi} - e^{\Delta\xi\psi}}, \tag{3.13}$$

where $\psi = 1 - f/f_0$.

The function (3.13) is substituted into (3.12) to give an equation which contains T only in the explicit form and two parameters ρ_2/ρ_1 and Δ:

$$\frac{e^{\xi\Delta/f_0} (e^{\Delta\xi\psi} - 1) + e^{\Delta\xi} - e^{\Delta\xi\psi}}{e^{\Delta\xi} - 1} = \left[\frac{e^{-\xi\Delta/f_0} (e^{\Delta\xi} - e^{\Delta\xi\psi}) + e^{\Delta\xi\psi} - 1}{e^{\Delta\xi\psi} - 1}\right]^{\frac{\rho_2}{\rho_1}}.$$

These parameters can be determined by requiring that the theoretical curve $n_l(T)/n_0$ passes through experimental points at two given temperatures T_1 and T_2. It is convenient to select these temperatures so that they are not too close to one another and sufficiently far from the flat parts of the dependence $n_l(T)/n_0$. We selected the temperatures $T_1 = 1.5$°K and $T_2 = 3$°K. We then obtained a system of two equations with two unknowns ρ_2/ρ_1 and Δ and we solved it by the branching method. The values of ρ_2/ρ_1 and Δ found in this way were used to plot the theoretical dependence $n_l(T)/n_0$ throughout the investigated temperature range. Figure 29 gives the results of measurements and the calculated temperature dependences of the density of localized electrons in three samples. The results indicate that the measured and calculated quantities agree quite satisfactorily. The relevant values of Δ, $E_0 = \Delta/f_0$, ρ_2, and ρ_1 are listed in Table 4.

It is clear from Table 4 that the calculated densities of states ρ_1 and ρ_2 are quite close to the values expected from a general discussion of the physical mechanisms (§ 4). The values

TABLE 4

Sample	Δ, °K	E_0, °K	$\rho_1 \cdot 10^{-31}$, erg^{-1} · cm^{-3}	$\rho_2 \cdot 10^{-32}$, erg^{-1} · cm^{-3}
7	5.6	7.8	1.2	6.4
13	3.1	6.1	3.1	11
15	1.4	4.4	6.6	16

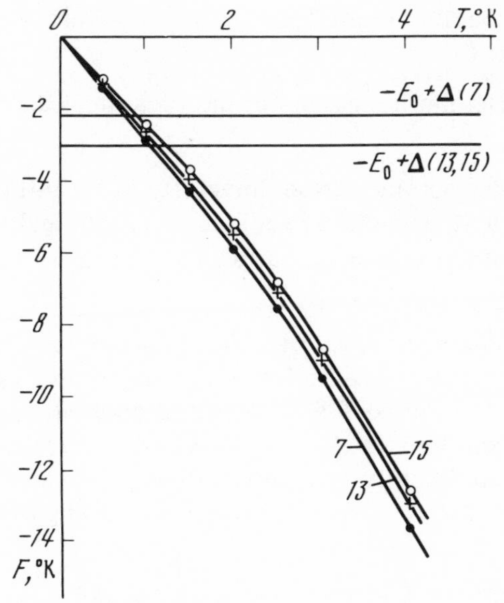

Fig. 32. Temperature dependences of the Fermi level.

of the width of the localized-states band Δ are of the order of the activation energy ε (§ 2a), i.e., they are of the order of the value governed by the slopes of the linear parts of the dependences $\ln(R_T/R_{5°K}) = f(1/T)$ in Fig. 25 but are slightly higher than the slopes. The values of Δ decrease with increasing number of electrons in the conduction band, which may be explained by an increase in the contribution of electrons to the screening of the potential wells, which reduces their depth.

Figure 32 gives the temperature dependences of the Fermi level calculated on the basis of Eq. (3.13). We selected the energy scale so that the Fermi level is F = 0 at T = 0. The electron gas is then degenerate and occupies the whole energy band $-E_0 \leq E \leq 0$ where the density of states is $\rho = \rho_1$. At higher temperatures the Fermi level descends and until this level reaches the value $-E_0 + \Delta$, the conduction electron density remains practically constant. Further shift of the Fermi level with heating increases the conduction electron density because of liberation of the localized electrons and it reaches saturation when the Fermi level becomes $F = -E_0$. At higher temperatures the electron gas becomes nondegenerate and the degree of degeneracy can be deduced from the value of the quantity $\gamma = (F + E_0 - \Delta)/T$ (Δ, E_0, and F are

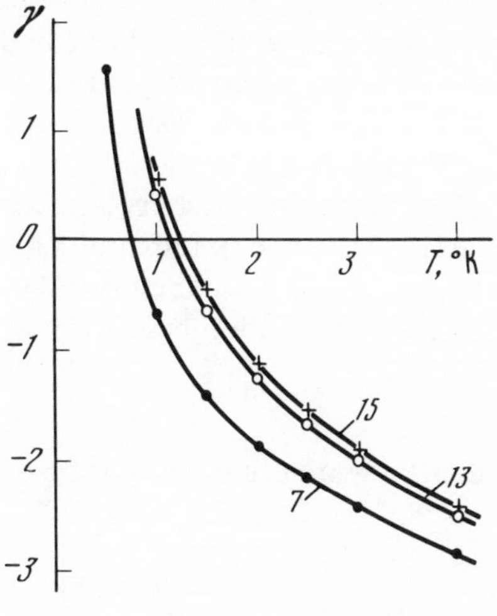

Fig. 33. Temperature dependences of the degeneracy factor $\gamma = (F + E_0 - \Delta)/T$.

all given in degrees Kelvin). The dependence of γ on T is plotted in Fig. 33. It is clear from this figure that the degeneracy disappears at $T \geq 1.5°K$.

We shall conclude this chapter by stressing the main features of the treatment adopted to deal with electrical conduction in our samples.

The temperature dependences of the resistivity obtained experimentally are similar to those reported by many authors for compensated semiconductors, such as Ge, InSb, and others (Chap. I, § 4), which may be regarded as evidence of the universal nature of the effects involved.

We shall assume that the dc conduction is of the band type, i.e., that it is due to free electrons and not due to hopping as assumed by most authors. The activation energy which appears at 5-1.5°K is due to a reduction in the number of free electrons as a result of cooling. This reduction may be explained by partial localization in the localized-states band adjoining the free-electron band. This makes it possible to explain consistently and quantitatively the experimental data obtained in the temperature range $0.5 \leqslant T \leqslant 300°K$ for strongly compensated gallium arsenide.

The general features of our energy scheme (Fig. 31) are similar to the model adopted for materials with a randomly distributed potential shown in Fig. 3 (Chap. I, § 4) because our step-like model of the density of states can be regarded as a rough description of any smooth variation of the density of states with energy. The available experimental data are insufficient to draw more specific conclusions on the energy dependence of the density of states.

A calculation based on the proposed model shows that an increase in the degree of compensation widens the localized band, whereas the Fermi level at $T \to 0$ is always located in the free-electron band. Thus, compensation does not give rise to hopping conduction at impurity concentrations such that the impurity band merges with the conduction band (Chap. I, § 3). This is easily understood because the total number of electrons in our samples is greater than the number of electrons which can be localized. The degree of compensation (up to 99%) is still insufficient for the Fermi level at $T \to 0$ to cross the localization boundary (in our case $E_c = -E_0 + \Delta$). Only further investigations can determine whether the predicted behavior of Δ and the Fermi level is common to all semiconductors with a sufficiently high concentration of the main impurity and with strong compensation. If we can show that a further increase in the degree of compensation changes the conduction mechanism to the hopping type in the limit $T \to 0$, this may be regarded as evidence that electrons are localized at impurity fluctuations. The nature of electron localization in compensated heavily doped materials is still not clear.

The investigations reported in the present chapter revealed also that the activation energy disappears when a critical electron density is reached. A strong dependence of ε (or Δ) on the electron density gives us grounds for assuming that this phenomenon is similar to the Mott transition (Chap. I, § 3). Our case differs because there is no qualitative jump – a transition from one type of conduction to another, which is characteristic of the insulator–metal transition. The screening of charged centers (groups of impurities) in our samples results only in a quantitative jump manifested by a steep change in the free-electron density. However, right down to the lowest temperatures the conduction in our model is due to electrons which remain free in the conduction band.

Appendix

A centrally symmetric potential well of depth U has a spectrum given by the equation [80]

$$\frac{\sigma}{\sqrt{\frac{U}{U_0} - \sigma^2}} = -\tan\frac{\pi}{2}\sigma,$$

TABLE 5

Well	U, meV	σ	E, meV	r_c/r_D
U_2	17	1.28	3	1.1
U_3	25.5	1.4	8,8	0.62
U_4	34	1.473	15,6	0.47
U_5	42.5	1.523	22,8	0.39

where U_0 is the minimum depth of the well which may capture an electron and which can be calculated from Eq. (3.6); E is a level in this well; U − E is the gap between the level and the bottom of the well;

$$\sigma=\sqrt{\frac{U}{U_0}-\frac{E}{U_0}} \text{ or } U-E=\sigma^2 U_0.$$

The radius of an electron state is

$$r_c=\hbar \Big/ \sqrt{2m^*E}=\frac{2}{\pi} r_D \sqrt{U_0/E}.$$

If U_0 = 8.5 meV and $r_D = 1.2 \cdot 10^{-6}$ cm, a calculation gives the parameters of the wells U_2-U_5 listed in Table 5.

CHAPTER IV

INDUCED "MOTT TRANSITION"

§ 1. Results of Photoconductivity Measurements

It is reported in Chap. III that there is a critical electron density in our samples amounting to $n_{0.cr} \approx 5 \cdot 10^{16}$ cm^{-3} (n_0 is the electron density at T = 300°K), which divides the samples into two groups. In samples with electron densities below the critical value we find that at low temperatures some of the conduction-band electrons are localized in wells which distort the potential relief. The temperature dependences of the electron density in such samples are characterized by an activation energy ε which is close to the width of the localized-states band Δ. When the free-electron density is increased from one sample to another, the values of ε and Δ decrease. At the critical density, electrons screen completely the potential wells with localized states and electron delocalization takes place. The activation energy of samples with electron densities above the critical value is $\varepsilon = 0$. This is analogous to the insulator−metal transition in a perfect crystal lattice caused by the electron screening of isolated charged centers and may take place more or less suddenly when the free-electron density is varied (Chap. I, § 1).

In the investigations described in Chap. III [69, 70] each electron density corresponds to a different sample. In the present chapter we shall report measurements of the free-electron density carried out by the present author on the same sample: the number of free electrons was varied by illumination [82]. In Chap. II, § 3 we described apparatus for measuring the photoconductivity of samples illuminated with ruby laser radiation of 1.78 eV energy, which was slightly greater than the forbidden band width of gallium arsenide.

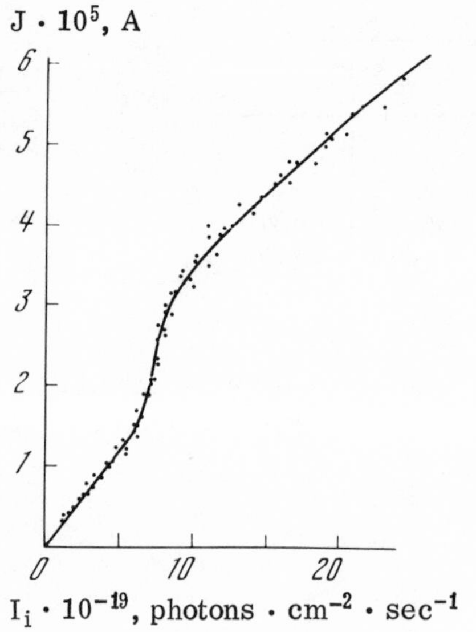

Fig. 34. Dependence of the photocurrent J on the illumination intensity I_i for sample 7.

The dependences of the photoconductivity on the illumination intensity at T = 1.6°K obtained using this apparatus will be considered in the present chapter in the specific case of samples 7 and 5, characterized by $n_{0.cr} > n_0 = n_l + n_c$, where n_l is the density of localized electrons at the measurement temperature and n is the density of free (conduction) electrons at the same temperature.

Since the duration of the photocurrent pulses was greater than the duration of the light pulses, we assumed that $t_i \leq \tau$, where t_i is the duration of illumination and τ is the nonequilibrium carrier lifetime. Therefore, the number of photons absorbed by a sample during one light pulse is $N_{ph} = I_i t_i S$, where I_i is the illumination intensity penetrating into the sample and S is the area of the illuminated surface.

As mentioned in Chap. II, the photocurrent at a given illumination intensity was taken to be the amplitude of the photocurrent pulse. Figures 34 and 35 give the results of measurements of the photocurrent J in samples 7 and 5 as a function of I_i. It is clear from these figures that the smooth curve $J = f(I_i)$ has an inflection at an intensity $I_{i.cr}$. This is shown more clearly in Figs. 36 and 37, which give the dependences of $J/N_{ph} = \alpha$ on N_{ph} for the same samples.

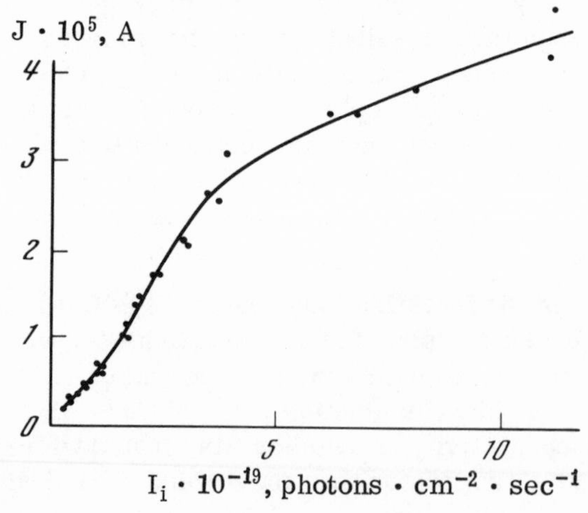

Fig. 35. Dependence of the photocurrent J on the illumination intensity I_i for sample 5.

TABLE 6

Sample	$J_{cr} \cdot 10^5$, A	$N_{ph.cr}$, photons	S, cm^2	$n_l \cdot 10^{-15}$, cm^{-3}	$n_c \cdot 10^{-15}$, cm^{-3}	μ_0, $cm^2 \cdot V^{-1} \cdot sec^{-1}$
7	2.1	$3.4 \cdot 10^{11}$	0.13	6	7	40
5	0.75	$5 \cdot 10^{10}$	0.065	4	4	80

Some of the initial data and the experimental results for these samples are given in Table 6, where n_c, n_l, and S have the same meaning as before; J_{cr} and $N_{ph.cr}$ are the critical values of the photocurrent and the number of absorbed photons at the point of inflection; μ_0 is the equilibrium electron density, i.e., the mobility of electrons in the absence of illumination at T = 1.6°K, whose value is taken from the experimental data given in Chap. III.

It is clear from Figs. 36 and 37 that the ratio $\alpha = J/N_{ph}$ before the inflection and far from it depends on N_{ph}. Hence, we may assume that in the investigated range of illumination intensities the photocarrier density is directly proportional to this intensity. Therefore, the inflection ("jump") of the current can only be due to the participation of additional carriers not generated by illumination.

§ 2. Calculation of the Photocurrent. Critical Illumination Intensity

In quantitative estimates we have to consider the components of the measured photocurrent J. At low illumination intensities, where the photocurrent rises smoothly with the illumination intensity, the value of J is governed by the number of electrons and holes generated by the incident light flux, and with the change in the equilibrium carrier density due to an increase in their mobility because of illumination: $J = J_e + J_h + J_c$.

The numbers of electrons and holes crossing in 1 sec a cross section dh of a photoconducting layer of depth d and width h (width of the sample) is $n_e dh\mu_e E + n_h dh\mu_h E + n_c dh(\mu_e - \mu_0)E$, where n_e and n_h are the densities of electrons and holes created by illumination; n_c is the density of equilibrium (conduction) electrons; μ_e and μ_h are the mobilities of photoelectrons and photoholes, respectively; μ_0 is the dark mobility of electrons. Since the carrier density is $n = N/Ldh$, where N is the total number of carriers in a photoconducting layer of length L, it follows that the current measured during a pulse is

$$J = \frac{eE}{L}(N_e\mu_e + N_h\mu_h + N_c\Delta\mu); \quad \Delta\mu = \mu_e - \mu_0.$$

If we assume that each photon creates a pair of carriers, we find that $N_e = N_h = N_{ph}$.

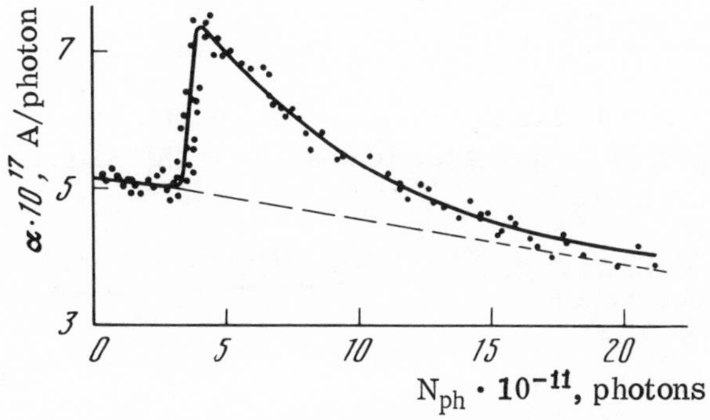

Fig. 36. Dependence of the ratio $\alpha = J/N_{ph}$ on N_{ph} for sample 7 at T = 1.6°K.

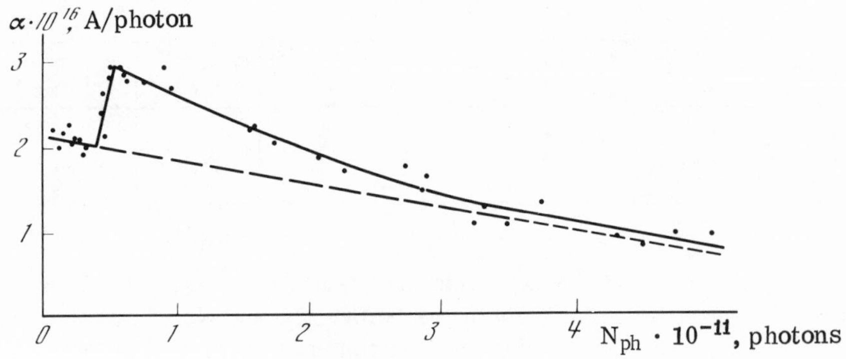

Fig. 37. Dependence of the ratio $\alpha = J/N_{ph}$ on N_{ph} for sample 5 at T = 1.6°K.

Since III-V compounds are characterized by $\mu_h \ll \mu_e$ [7], we can ignore the contribution of the hole component of the current.

We shall also assume that the influence of the surface recombination is negligible. In our samples such recombination is probably slight because of the etching in a polishing solution before the measurements (Chap. II, § 1).

The jump (or more correctly, an inflection) of the photocurrent observed at the critical intensity $N_{ph.cr}$ will be assumed to be due to the delocalization of electrons from the potential wells. Under photoexcitation conditions and in the case of delocalization, the photocurrent is

$$J = \frac{eE}{L}\left(N_{ph}\mu_e + N\mu_e + N_c\,\Delta\mu\right),$$

where N is the number of delocalized electrons.

In the limit $N_{ph} \to 0$, the photocurrent becomes $J \to 0$ because $N = 0$ and $\Delta\mu \to 0$. Then,

$$\alpha(N_{ph}) = \frac{J}{N_{ph}} = \frac{e\mu_e E}{L}\left(1 + \frac{N}{N_{ph}} + \frac{N_c\,\Delta\mu}{N_{ph}\,\mu_e}\right).$$

The experimental dependences $\alpha(N_{ph})$ are represented by points in Figs. 36 and 37. If $N = 0$, we find that

$$\alpha = \alpha_0 = \frac{e\mu_e E}{L}\left(1 + \frac{N_c\,\Delta\mu}{N_{ph}\,\mu_e}\right). \qquad (4.1)$$

It is clear from Figs. 36 and 37 that α_0 decreases monotonically with increasing N_{ph} but the rate of this change is so slow that in the narrow range of the intensities in the region of the jump in the current we can ignore the change in α_0.

In the region of the jump (and after it), the value of N is equal to N_l, which is the total number of delocalized electrons, so that the photocurrent is

$$J = \frac{e\mu_e E}{L}\left(N_{ph} + N_l + N_c\,\frac{\Delta\mu}{\mu_e}\right), \qquad (4.2)$$

and the ratio defined above is

$$\alpha=\frac{J}{N_{\mathrm{ph}}}=\frac{e\mu_e E}{L}\left(1+\frac{N_l}{N_{\mathrm{ph}}}+\frac{N_c}{N_{\mathrm{ph}}}\frac{\Delta\mu}{\mu_e}\right). \tag{4.3}$$

It follows from Eqs. (4.1) and (4.3) that

$$\frac{\alpha-\alpha_0}{\alpha_0}=\frac{N_l}{N_{\mathrm{ph}}+N_c\dfrac{\Delta\mu}{\mu_e}}. \tag{4.4}$$

Since N = ndS, it follows that the system (4.2) and (4.4) describing the jump of the current has only two unknowns: d and μ_e. Some experimental data and the results of solution of the system (4.2) and (4.4) are given in Table 7. The effective electron temperatures T_e corresponding to the hot-electron mobility μ_e are taken from the experimental results described in Chap. III.

It follows from Eq. (4.4) that

$$\alpha_0=\alpha\left(1+\frac{N_l}{N_{\mathrm{ph}}+N_c\dfrac{\Delta\mu}{\mu_e}}\right)^{-1}. \tag{4.5}$$

At high values of N_{ph} and, consequently, high values of μ_e we have $\Delta\mu\to\mu_e$, because $\mu_0 \ll \mu_e$ and then

$$\alpha_0\to\alpha\left(1+\frac{N_l}{N_{\mathrm{ph}}+N_c}\right)^{-1}=\alpha\left(1+\frac{n_l}{n_{\mathrm{ph}}+n_c}\right)^{-1}\to\alpha.$$

Therefore, Eq. (4.5) can be used to calculate readily the value of α_0 at high intensities N_{ph} and the results obtained in this way are represented by the parts of the lines composed of short dashes. If we know the change in α_0 before the jump and the dependence of α_0 on N_{ph} in the range of high values of N_{ph}, we can extrapolate the dependence $\alpha = f(N_{\mathrm{ph}})$ to moderate values of N_{ph}, as shown in Figs. 36 and 37 by lines composed of longer dashes.

§ 3. Critical Electron Density

It will be shown later (§ 4) that overheating of the lattice is $\Delta T < 0.7$°K and, therefore, its influence on the electron delocalization can be ignored. It follows from the results obtained and from their interpretation that localized electrons are not liberated directly during a pulse by photons or by hot electrons.

Clearly, the only reason for the jump of the current observed under our conditions can be a phenomenon analogous to the Mott transition. According to the Mott criterion (1.1), a sudden transition from an insulating to a metallic state is due to the screening of charged

TABLE 7

Sample	$\frac{\alpha-\alpha_0}{\alpha_0}$	$d\cdot 10^4$, cm	μ_0, $cm^2\cdot V^{-1}\cdot sec^{-1}$	μ_e, $cm^2\cdot V^{-1}\cdot sec^{-1}$	T_e, °K
7	0.43	2	40	130	15.6
5	0.45	1.6	80	190	20

TABLE 8

Sample	$n_l \cdot 10^{-15}$, cm^{-3}	$n_c \cdot 10^{-15}$, cm^{-3}	n_0, cm^{-3}	$n_{ph.cr} \cdot 10^{-16}$, cm^{-3}	$n_{cr} \cdot 10^{-16}$, cm^{-3}
7	6	7	$1.3 \cdot 10^{16}$	1.0	2.3
5	4	4	$8 \cdot 10^{15}$	0.7	1.5

centers by free electrons. The Mott criterion relates the critical density of free electrons to the Bohr radius of a charged center by Eq. (1.1):

$$n_{cr}^{1/3} a_0 \approx 0.25.$$

The conditions for a Mott jump in our case are not identical with those for which Eq. (1.1) is calculated. For our samples the Maxwellian relaxation time is $\tau_M \sim 10^{-11}$ sec. Therefore, the illumination pulse duration $t_i = 4 \cdot 10^{-8}$ sec is quite sufficient for the screening and, instead of the Bohr radius, the characteristic length can be taken to be the average width of a potential well (Chap. III, § 4), which is 10^{-6} cm. It then follows from the Mott criterion that the critical density should be $n_{cr} = 1.6 \cdot 10^{16}$ cm^{-3}.

If we know the experimental values of $N_{ph.cr}$ and the calculated thickness of the photoconducting layer d, we can readily determine the photoelectron density at which delocalization takes place, i.e., at which an analog of the Mott transition occurs:

$$n_{ph.cr} = N_{ph.cr}/dS.$$

The total density of electrons causing delocalization is equal to $n_{cr} = n_0 + n_{ph.cr}$, where n_0 is the total density of electrons in the conduction band in darkness. The results of calculations of $n_{ph.cr}$ and n_{cr} are given in Table 8.

It is clear from Table 8 that the value of n_{cr} deduced from the jump of the current is close to the calculated value obtained from the Mott criterion. However, the closeness between the measured and calculated values should not be taken too seriously. Mott himself described Eq. (1.1) as only an approximate relationship [12]. We may assume that the agreement with his estimates only qualitatively confirms the assumptions underlying the calculations and the assumed physical model. Clearly, the delocalization of electrons caused by illumination is due to photoelectrons and photoholes. When the illumination intensity reaches a value at which the electron and hole densities are sufficient to screen the potential brarriers, localized electrons become free almost instantaneously.

§ 4. Estimate of Overheating of the Lattice as a Result of Illumination

It is mentioned in § 3 that the delocalization of electrons in our experiments was not due to the heating of the crystal lattice by the incident light. In the present section we shall give approximate estimates of the lattice heating and consider qualitatively the phenomena which accompany the appearance of nonequilibrium phonons in a sample. The nature of these phenomena depends on the relationship between the phonon mean free path, size of the sample, and duration of light pulses.

Investigations of the motion of nonequilibrium phonons, usually created by a short thermal pulse, in solids have become popular [83]. A thermal pulse creates a phonon packet traveling from one end of a crystal to another. Before the completion of the phonon relaxation processes, the phonon transport occurs in the ballistic regime [84]. This regime represents simply the propagation of sound, described by a wave equation, and the phonon velocity is equal to the velocity of sound u in a crystal. The energy carried by phonons moving in the ballistic regime is not the thermal energy, i.e., it is not the energy of isotropic vibrations of the crystal lattice atoms, until the end of the relaxation of nonequilibrium phonons.

In a thin and pure sample such phonons reach the other end of the sample before they relax. In thicker samples the phonons do not reach the other end in the relaxation time and the sample temperature rises but this rise depends on the relationship between the duration of the light pulse and the relaxation time. If nonequilibrium phonons transform to equilibrium particles during a light pulse, an estimate of the heating of a crystal should allow for the diffusion of heat from the illuminated surface to the dark part of the sample.

The total relaxation time is given by the relationship [85] $\tau^{-1} = \tau_b^{-1} + \tau_I^{-1} + \tau_p^{-1}$, where τ_b is the relaxation time due to the interaction with boundaries; τ_I is the relaxation time due to the interaction with impurities; τ_p is the relaxation time due to phonon–phonon collisions.

We shall estimate the relaxation time τ of our samples using the results reported in [85] for gallium arsenide with an impurity concentration $\sim 10^{18}$ cm^{-3}. For the dimensions of our samples these estimates give $\tau_b \sim 10^{-6}$ sec, $\tau_I \sim 10^{-10}$ sec, $\tau_p \sim 10^{-6}$ sec, which shows that the scattering by impurities should play the dominant role in the relaxation in our samples. These estimates show also that in our samples heat should be transported by diffusion because the mean free path of phonons is $l_p = u\tau \approx 5 \cdot 10^5 \cdot 10^{-10} = 5 \cdot 10^{-5}$ cm, which is much less than the thickness of our samples $\sim 10^{-1}$ cm, and because the phonon relaxation time is much less than the duration of the light pulses $\tau \ll t_i = 4 \cdot 10^{-8}$ sec.

We shall now estimate the overheating of a sample on the assumption that heat is transported by diffusion. If the heating due to the absorption of ruby laser radiation is an adiabatic process, the power delivered by a laser pulse is converted entirely into heat which raises the temperature of a sample from T_1 to T_2:

$$Q_i = \int_{T_1}^{T_2} C(T)\, m\, dT,$$

where m is the mass of the heated layer, whose thickness is equal to the sum of the thickness of the photoconducting layer and of the thermal diffusion length; C(T) is the specific heat of n-type GaAs, which is a function of temperature.

The thermal diffusion length is [86]

$$\lambda \approx \sqrt{k(T)\, t_i / \rho C(T)},$$

where k(T) is the thermal conductivity and ρ is the density of GaAs.

We shall now estimate the thermal diffusion length at T $\approx$ 2°K. The low-temperature specific heat of pure gallium arsenide [87] is

$$C'(T) = \sum_{n=0}^{10} a_{2n+1} T^{2n+1} \approx a_1 T + a_3 T^3 \approx a_3 T^3 \text{ J} \cdot \text{g-atom}^{-1} \cdot \text{deg}^{-1},$$

because $a_1 = (0.8 \pm 0.8) \cdot 10^{-6}$ and $a_3 = (47.52 \pm 0.9) \cdot 10^{-6}$. Bearing in mind that one g-atom of GaAs is 72.3 g [88], we find that

$$C(T) = 0.657 \cdot T^3 \cdot 10^{-6} \text{ J} \cdot \text{g}^{-1} \cdot \text{deg}^{-1}.$$

The thermal conductivity k(T) is a function of temperature and it varies strongly with the impurity concentration in a sample. We shall estimate the thermal conductivity of our samples employing the data reported in [85] for GaAs with an impurity concentration similar to ours. According to [85], the thermal conductivity of a sample with an impurity concentration $N_i \sim 10^{18}$ cm^{-3} is described by the relationship $k(T) = 0.093 \cdot T^{2.5}$ J·cm^{-1}·deg^{-1}·sec^{-1} and at $T \approx 2$°K the conductivity is $k \approx 0.53$ J·cm^{-1}·deg^{-1}·sec^{-1}.

The thermal diffusion length $\lambda \approx 5 \cdot 10^{-2}$ cm calculated for $T \approx 2$°K is slightly longer than the distance traveled by nonequilibrium phonons during an illumination pulse $l = ut_i \approx 2 \cdot 10^{-2}$ cm. The concept of the thermal diffusion length is meaningful only as long as the diffusion coefficient $k(T)/\rho C(T)$ depends weakly on temperature. In our case, we have $k(T)/\rho C(T) \propto T^{0.5}$ and, therefore, our estimates are only valid to a degree to which we can assume this quantity to be independent of temperature. Moreover, the diffusion processes are limited in time because the rate of propagation of heat becomes $\lambda/t \rightarrow \infty$ in the limit $t \rightarrow 0$. The diffusion approximation can only be used if $\lambda/t_i > u$. Therefore, we shall assume that the depth of propagation of heat during a light pulse is $l \approx 2 \cdot 10^{-2}$ cm.

Since the thickness of a photoconducting layer is $d \sim 10^{-4}$ cm $\ll l$, we can take l to be the thickness of the heated layer. Then, the mass of the heated layer is $m = \rho S l$ (g). The photon energy is $h\nu = 1.78$ eV $= 1.78 \cdot 1.6 \cdot 10^{-12} \cdot 10^{-7}$ J $= 2.85 \cdot 10^{-19}$ J. During a light pulse the number of photons reaching a sample is $I_i S t_i$ so that $Q = 2.85 \cdot 10^{-19} I_i S t_i$ (J).

We can determine the temperature T_2 from

$$2.85 \cdot 10^{-19}\, I_i S t_i = S \rho l \int_{T_1}^{T_2} C(T)\, dT,$$

and the solution of this equation gives $T_2 = 2.35$°K.

Thus, if we assume that the heating is adiabatic, we find that during a light pulse a sample absorbs $Q_i \sim 10^{-7}$ J and the part of a crystal where the photocurrent appears is heated 1.6 to 2.3°K, i.e., the temperature rise ΔT is 0.7°K. However, this should occur in vacuum and in liquid helium the process may not be adiabatic at all. Heat transfer in liquid helium (at temperatures $T_1 < 2.2$°K) involves the propagation of temperature waves (second sound). The velocity of second sound is a function of temperature and it amounts to $2 \cdot 10^3$ cm/sec at T = 1.6°K [89]. Since the specific heat of helium is relatively high ($C_v \approx 7$ J·mole^{-1}·deg^{-1}), the presence of temperature waves prevents the establishment of any temperature gradient in the liquid itself [90]. If all the heat evolved during a pulse is transferred to liquid helium, the temperature of the sample should not rise at all: $\Delta T = 0$. However, the transport of heat from a solid to liquid helium is not limited to the thermal conductivity of bulk liquid helium. At the boundary between a solid and liquid helium there is a Kapitsa temperature jump, which is a temperature barrier $\Delta T = KW$, where K depends on the properties of the solid and the state of its surface and W is the heat flux from the solid to the surrounding liquid helium. Unfortunately, there is no information on the Kapitsa jump in the case of GaAs but it is likely that some of the power dissipated in a sample is transferred to the surrounding medium and this reduces the temperature rise in the sample, i.e., we can assume that $\Delta T < 0.7$°K.

Measurements of the temperature rise in germanium illuminated with xenon lamp flashes of 3 μsec duration are reported in [91]. When the average carrier density in a sample was $\sim 10^{16}$ cm^{-3}, the temperature rise was $\Delta T \le 0.2$°K if the initial temperature was T = 2°K. It was stressed in [91] that this temperature rise, measured with a superconducting bolometer on the dark side of a sample, was somewhat less than that calculated without allowance for the loss of heat to the surrounding helium. Since many properties of gallium arsenide (for example, the specific heat) are similar to those of germanium, we may assume that this is also true of our case. Similar measurements of the rise of temperature in germanium were also carried out at the Lebedev Physics Institute [92].

§ 5. Electron Gas Temperature

The values of the mobility μ_e listed in Table 7 indicate that electrons participating in a photocurrent are hot and that their effective temperature – deduced from the temperature dependence of the mobility (Chap. III) – may reach 15-20°K.

We shall now consider the processes which occur in the electron gas in the course of absorption of light. Each photon creates an electron and a hole of energy which is distributed between them approximately in inverse proportion to their masses. Optically created electrons receive an energy which is large compared with the energy of equilibrium electrons.

Nonequilibrium electrons lose their energy in several stages.

If the electron density is higher than the critical value discussed earlier (this value depends on the material, mechanism of the interaction with the lattice, and temperature), then before the beginning of cooling (by the emission of optical phonons) a thermodynamic equilibrium is established in the electron gas by electron–electron collisions.

At T = 4.2°K even for $n > n_{cr} \sim 10^{11}$ cm^{-3} the exchange of energy by electron–electron collisions in gallium arsenide becomes faster than in collisions with optical and acoustic phonons [75, 93, 94]. Since the electron density in our samples is higher than the critical value (it is of the order of 10^{16} cm^{-3}), a thermodynamic equilibrium is established before the emission of optical phonons and this happens in a time interval of the order of [93, 95]

$$\tau_{ee} \sim \frac{\varepsilon^{3/2} (m_e^*)^{1/2} \varkappa^2}{\pi e^4 n},$$

where $\varkappa$ is the permittivity; ε is the electron energy; n is the electron density.

We shall estimate this time interval by determining the energy of electrons after the absorption of light. Each photon of $h\nu - E_g = 1.78 - 1.53 = 0.25$ eV. Of this energy an electron receives $0.9 \cdot 0.25 = 0.225$ eV because $m_h^*/m_e^* \approx 10$ [71]. For $\varepsilon = 0.225$ eV the relaxation time of the electron gas is $\tau_{ee} \sim 10^{-10}$ sec, and, therefore, we may assume that at each moment during a light pulse of $4 \cdot 10^{-8}$ sec duration the electron gas is not a mixture of electrons of different energies but is a system with a steady-state temperature.

If this temperature is high, then the emission of optical and acoustic phonons begins. Cooling by the emission of optical phonons is approximately 100 times faster than by the emission of acoustic phonons [96]. The slowest process in the cooling of the electron gas is that involving losses due to the interaction with acoustic phonons. The energy lost by hot carriers interacting with acoustic phonons is calculated in [95]. According to these calculations, the cooling time of carriers from an initial energy ε to an energy $\bar{\varepsilon}$ is

$$t_{ac} = \frac{2}{a}\left(\frac{1}{\bar{\varepsilon}^{1/2}} - \frac{1}{\varepsilon^{1/2}}\right), \quad (4.6)$$

$$a = \frac{2\,(m^*)^{1/2} u^2}{l_{ac} kT},$$

where T is the temperature of the sample; l_{ac} is the mean free path of electrons scattered by acoustic phonons; m* is the effective mass of an electron; u is the velocity of sound.

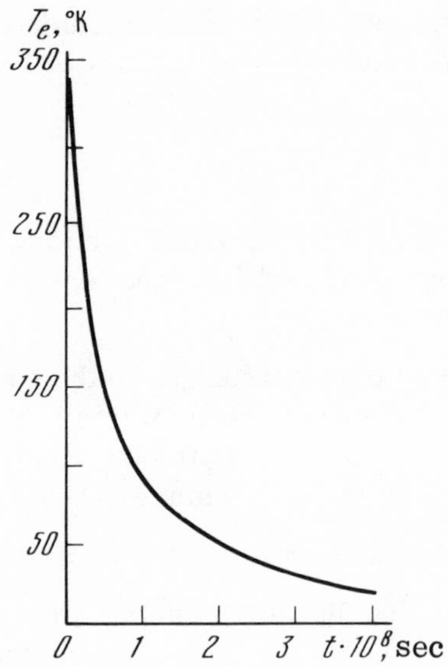

Fig. 38. Cooling of the electron gas in accordance with Eq. (4.6) in the absence of free holes.

It is clear from the above expression that the carrier cooling time depends strongly on the effective mass and, therefore, cooling of an electron–hole plasma depends on the proportion of the holes captured by traps and those remaining free.

For electrons in gallium arsenide we have $m^* = 0.07m_0$, $u = 4.8 \cdot 10^5$ cm/sec [97] so that the product $l_{ac}T$ is $6.7 \cdot 10^{-2}$ cm/deg (see Chap. V and [98]) and, therefore, $a_e = 5.6 \cdot 10^{14}$ $erg^{-1/2} \cdot sec$.

If we use the data reported in [71, 98], we find that in the case of holes we have $a_h \sim 10^{17}$ $erg^{-1/2} \cdot sec$, i.e., in this case the cooling is much faster. For the time being we shall assume our samples to not contain free holes or very few of them (this is confirmed by estimates given later) and we shall follow the cooling of the electron gas heated to a temperature corresponding to the energy of one optical phonon. The Debye temperature of gallium arsenide is $T_D = 345°K$ [87] and, therefore, $\hbar\omega_0 = kT_D = 0.03$ eV. Cooling of electrons calculated using Eq. (4.6) and the constant $a_e = 5.6 \cdot 10^{14}$ $erg^{-1/2} \cdot sec$ is shown in Fig. 38. During a light pulse each photon produces new high-energy nonequilibrium electrons and at the same time the electron gas cools because its temperature is higher than the lattice temperature, as manifested by the higher electron mobility in our samples.

Let us now consider how the cooling process may be affected by the presence of free holes. The time needed for the establishment of a thermodynamic equilibrium between electrons and holes, i.e., the energy relaxation time of an electron–hole plasma, is defined as the momentum relaxation time multiplied by the ratio of the effective masses of a hole and an electron [99]. For the highest electron energies the time to the onset of cooling is $\tau_{he} \sim \tau_{ee} \cdot 10 \sim 10^{-9}$ sec. As the electron energy decreases, this time becomes shorter because of reduction in τ_{ee}, so that for $\varepsilon \approx 0.03$ eV, the electron–hole plasma relaxation time is $\tau_{he} \sim 10^{-10}$ sec.

Since cooling even by interaction with acoustic phonons in the presence of holes is fast because of the large value of a_h and because $\tau_{he} \ll t_i$, in this case a sample should not contain any hot electrons. The values of the mobility μ_e corresponding to an electron temperature $T_e \approx 20°K$ indicate that holes in our samples may be captured by traps.

Thus, the establishment of the temperature of the injected electrons is a complex process which depends on many factors, and it is difficult to calculate the electron temperature.

We shall end this chapter by noting that the above explanation of the experimental results is valid even when the nonequilibrium carrier lifetime τ is less than t_i but then the electron density $n_{ph.cr}$ is less.

CHAPTER V

HEATING OF ELECTRONS IN STRONG ELECTRIC FIELDS

The physical model developed in Chap. III is based on the assumption that conduction in compensated gallium arsenide is due to free electrons throughout the investigated temperature range $0.5 \lessdot T \lessdot 300°K$. However, these free electrons move in a conduction band with a strongly distorted potential relief, which gives rise to localized states in this band. In the experiments [69, 70] described in Chap. III the electrical conductivity was measured at low temperatures in a weak electric field $E \leq 10^{-2}$ V/cm, when the temperature of electrons was practically equal to the lattice temperature. It seemed desirable to extend these measurements to the range of stronger electric fields [100].

§ 1. Current – Voltage Characteristics and the Field Dependence of the Electron Mobility

Typical dependences of the current density j on the field E obtained at various temperatures are plotted in Figs. 39 and 40. It follows that at helium temperatures in fields $E < 2$ V/cm the current density is practically proportional to E but in fields $E > 10$ V/cm it is proportional to E^2. In fields $E > 30$ V/cm some samples exhibit a dependence which is even stronger than $j \propto E^2$, but others exhibit a weaker dependence; however, in all cases it remains close to $j \propto E^2$.

Deviations of the dependence $j = f(E)$ from linearity in fields $E > 10$ V/cm may be due to an increase in the number of carriers or an increase in their mobility. Measurements of the Hall coefficient at $T = 4.2°K$ in the pulse regime using the circuit described in Chap. II, § 4

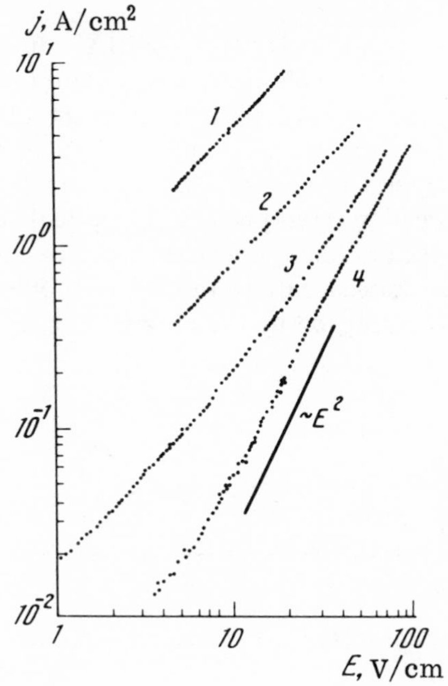

Fig. 39. Dependences of the current density j on the electric field E applied to sample 1 at various temperatures T (°K): 1) 290; 2) 77; 3) 20; 4) 1.8.

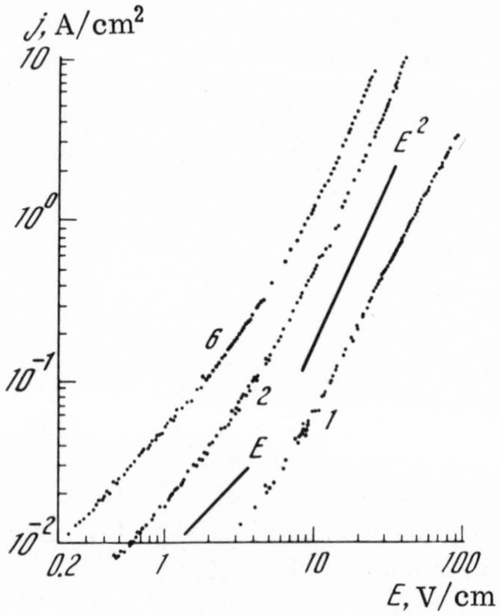

Fig. 40. Dependences of the current density j on the electric field E applied at T = 1.8°K to three samples with different electron densities.

indicated that, within the limits of an experimental error of ±10%, the carrier density remained constant throughout the investigated range of electric fields. The electron density was calculated from the experimental data assuming that the Hall factor was r = 1, as indicated by the results reported in Chap. III, § 1b. Figures 41 and 42 give the values of the electron density obtained for samples 2 and 6, respectively, as a function of the electric field. It is clear from these results that there is no rise in the number of electrons with increasing field. This may be due to the fact that at T = 4.2°K the majority of electrons in our samples is already delocalized and, within the limits of the experimental error, there is no change in the number of carriers.

Since the electron density can be regarded as constant, it follows that the rise of the electrical conductivity with the field is due to an increase in the electron mobility.

§ 2. Relaxation of the Electron Energy

It is shown in Chap III, § 2b that the temperature dependence of the mobility μ in our samples observed in the range T < 25°K may be explained by the dipole scattering. In this case we have $\mu \propto T_e^{1/2}$, where T_e is the electron temperature. The energy acquired from the electric field is lost by electrons in the interaction with the lattice vibrations because in collisions with impurities or groups of impurities the energy lost by electrons is negligible due to the large mass of the impurities. Moreover, the energy transferred by electrons to the optical lattice vibrations is negligible if the electron temperature is not too high compared with the lattice temperature T = 4.2°K. Thus, an electron may transfer energy only to the acoustic vibrations of the lattice either by scattering on the deformation potential, or by the piezoelectric scatter-

Fig. 41. Dependences of the electron density n on the electric field E applied to sample 2.

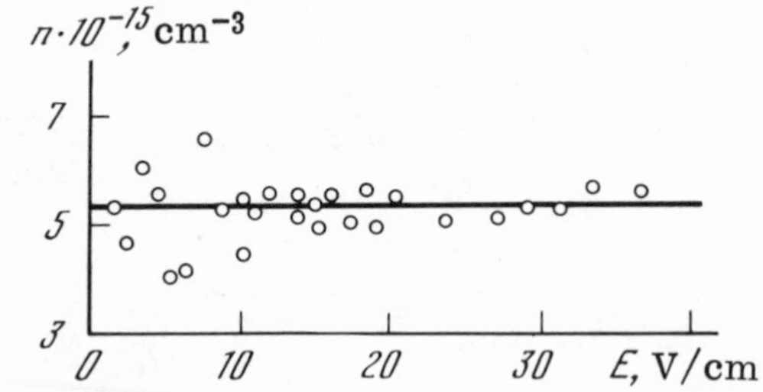

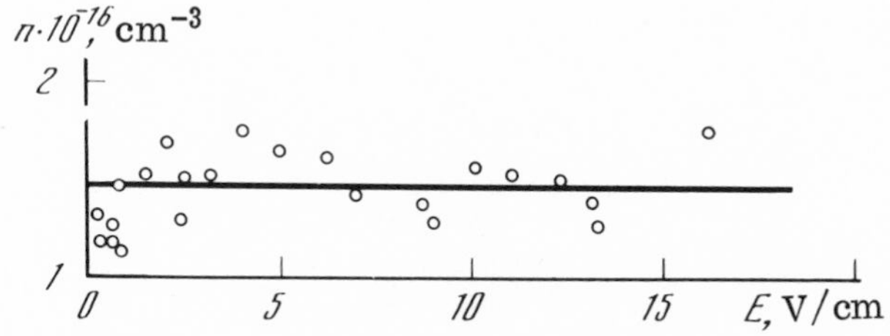

Fig. 42. Dependence of the electron density n on the electric field E applied to sample 6.

ing, which is due to the electric polarization that appears in a polar crystal as a result of lattice vibrations [101]. To the author's knowledge, the question of which of the last two types of scattering predominates has not yet been resolved. Ehrenreich [71] assumes that the importance of the piezoelectric scattering in III-V compounds has not been established because the piezoelectric constants of these materials are not known. Oliver [102] assumed that the electron momentum relaxation was due to the scattering by ionized impurities and calculated that at electron temperatures below 30°K the electron energy losses are mainly due to the piezoelectric scattering, but at temperatures above they are due to the scattering by the optical vibrations. Zylbersztejn [103] calculated that in InSb, which is also partly a polar crystal, the scattering of electrons by the deformation potential is important at helium temperatures. However, all these results should be regarded as preliminary [101] because of the contradictions between them.

§ 3. Calculation of the Effective Electron Temperature

If the electron energy relaxation is due to the scattering by the deformation potential and the electron momentum relaxation is due to the elastic scattering by impurities or groups of impurities, the electron distribution function in a homogeneous electric field E can be represented, as demonstrated by Chuenkov [104], by

$$f(\varepsilon, \theta) = f_0(\varepsilon) + f_1(\varepsilon)\cos\theta, \tag{5.1}$$

where $f_0(\varepsilon) \gg f_1(\varepsilon)$.

Here, ε is the electron energy; θ is the angle between the electron velocity and the direction of the electric field E; and the functions in the above expression are given by

$$f_0(\varepsilon) = C\exp\left[-\int\frac{d\varepsilon}{kT\left(1+\frac{e^2 l_{ac} l(\varepsilon) E^2}{6m^* u^2 \varepsilon}\right)}\right], \tag{5.2}$$

$$f_1(\varepsilon) = -eEl(\varepsilon)\frac{df_0(\varepsilon)}{d\varepsilon}, \tag{5.3}$$

where C is a normalization constant; k is the Boltzmann constant; T is the temperature of the investigated crystal; e is the electron charge; m* is the effective electron mass; u is the velocity of sound; l_{ac} is the mean free path of electrons scattered by acoustic phonons; $l(\varepsilon)$ is the mean free path for the main type of scattering.

If the electron momentum relaxation is mainly due to the scattering by dipoles, then

$$l_d(\varepsilon) = a\varepsilon, \tag{5.4}$$

where a is a constant.

Substituting Eq. (5.4) into Eq. (5.2), we obtain

$$f_0(\varepsilon) = C \exp\left(-\frac{\varepsilon}{kT_e}\right), \tag{5.5}$$

where

$$T_e = T\left[1 + \frac{e^2 l_{ac} l_d (kT) E^2}{6m^* u^2 kT}\right] \tag{5.6}$$

is the effective temperature of electrons; l_d(kT) is the mean free path of electrons scattered by dipoles taken at ε = kT.

The last function can be compared with the experimental results by replacing the mean free paths of electrons l_{ac} and l_d in Eq. (5.6) by the corresponding electron mobilities μ_{ac} and μ_d.

The electron mobility can be calculated from

$$\mu(E) = \frac{1}{E} \frac{\int_0^\infty \int_0^\pi v \cos\theta f(\varepsilon, \theta) \rho(\varepsilon) \, d\varepsilon \sin\theta \, d\theta}{\int_0^\infty \int_0^\pi f(\varepsilon, \theta) \rho(\varepsilon) \, d\varepsilon \sin\theta \, d\theta}, \tag{5.7}$$

where $\rho(\varepsilon)$ is the density of states and v is the electron velocity.

Equations (5.1)-(5.7) can be used to find the field dependence of the electron mobility in the dipole scattering case:

$$\mu_d(E) = \frac{4}{3}\sqrt{\frac{2}{\pi}} \frac{e l_d(T)}{\sqrt{m^* kT}} \sqrt{kT_e} = \frac{4}{3}\sqrt{\frac{2}{\pi}} \frac{e l_d(T)}{\sqrt{m^* kT}} \sqrt{\frac{T_e}{T}} = \mu_0 \sqrt{\frac{T_e}{T}}, \tag{5.8}$$

where μ_0 is the mobility in the limit E → 0 at a given crystal temperature T, i.e., for the dipole scattering when T_e = T:

$$\mu_0 = \frac{4}{3}\sqrt{\frac{2}{\pi}} \frac{e l_d(T)}{\sqrt{m^* kT}}. \tag{5.9}$$

For the scattering by acoustic phonons, we have

$$\mu_{ac} = \frac{2}{3}\sqrt{\frac{2}{\pi}} \frac{e l_{ac}(T)}{\sqrt{m^* kT}}. \tag{5.10}$$

Using Eqs. (5.9) and (5.10), we find from Eq. (5.6) that

$$T_e = T\left(1 + \frac{3\pi}{32u^2} \mu_{ac}\mu_0 E^2\right) = T + \zeta E^2, \tag{5.11}$$

where

$$\zeta = \frac{3\pi}{32u^2} \mu_{ac}\mu_0 T. \tag{5.12}$$

The mobility μ_0 can be assumed to be the experimentally obtained value. The mobility μ_{ac} can be calculated using the Bardeen–Shockley relationship [98]:

$$\mu_{ac} = \frac{2\sqrt{2\pi} e \hbar^4 c_{\|}}{3 (m^*)^{5/2} (kT)^{3/2} E_1^2},$$

where $c_{\|}$ is the elastic constant of gallium arsenide equal to ρu^2; ρ is the density of the crystal; E_1 is the deformation potential constant of the conduction band of gallium arsenide.

Calculations were carried out assuming that the elastic constant was $c_{\|} = 1.23 \cdot 10^{12}$ dyn/cm^2 [97], the effective carrier mass was $m^* = 0.07m_0$ [97], the deformation potential constant was $E_1 = -7$ eV [71], and the velocity of sound was $u = 4.8 \cdot 10^5$ cm/sec [97]. Then, it was found that at T = 4.2°K the mobility should be $\mu_{ac} = 7 \cdot 10^7$ cm$^2 \cdot$V$^{-1} \cdot$sec^{-1} and, consequently,

$$\zeta_{theor} = 3.8 \cdot 10^{-4} \cdot \mu_0 \ \text{deg} \cdot \text{sec} \cdot \text{V}^{-1}. \tag{5.13}$$

In particular, for sample 2 with the experimental mobility $\mu_0 = 90$ cm$^2 \cdot$V$^{-1} \cdot$sec^{-1}, the relationship (5.13) indicated that $\zeta_{theor} = 0.034$ deg$\cdot$cm$^2 \cdot$V^{-2}.

We shall compare the dependence of the mobility on the electric field E with the experimental data. It follows from Eqs. (4.8) and (5.11) that this dependence can be represented in the form

$$\mu = \mu_0 \sqrt{1 + \frac{\zeta}{T} E^2}. \tag{5.14}$$

The results of measurements on sample 2 are represented by points in Fig. 43 and the continuous curve is calculated on the basis of Eq. (5.14) assuming that $\zeta_{exp} = 0.045$ deg$\cdot$cm$^2 \cdot$V^{-2}. The agreement between the theoretical and experimental values ζ_{theor} and ζ_{exp} can be regarded as good for a simplified calculation which allows only for the scattering by dipoles and for the energy losses due to the interaction with the deformation potential.

The field dependence of the effective electron temperature $T = f(E^2)$ calculated in accordance with Eq. (5.11) for $\zeta_{theor} = 0.034$ deg$\cdot$cm$^2 \cdot$V^{-2} is represented by the dashed line in Fig. 44.

A field dependence of the effective electron temperature similar to Eq. (5.11) can be derived from the energy balance condition by equating the power $e\mu E^2$ acquired by a carrier from a field E to the average rate of energy loss by carriers as a result of scattering by acoustic phonons, calculated using the perturbation theory [101]:

$$e\mu E^2 = \left(\frac{d\varepsilon}{dt}\right)_{ac} = \frac{8\sqrt{2}E_1^2 (m^*)^{5/2} (kT_e)^{3/2}}{\pi^{3/2}\hbar^4 \rho}\left(1 - \frac{T}{T_e}\right). \tag{5.15}$$

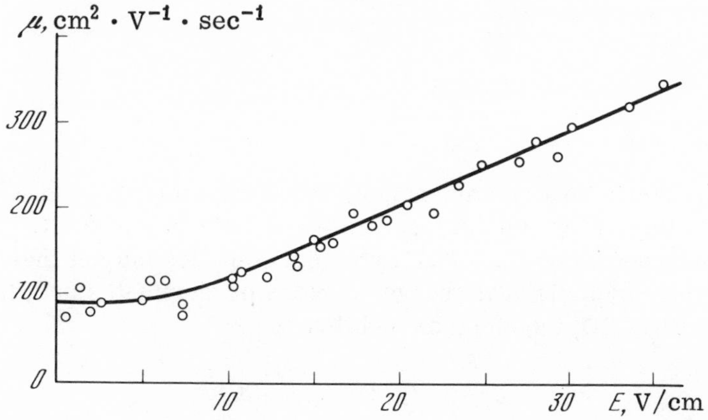

Fig. 43. Dependence of the electron mobility μ on the electric field E applied to sample 2.

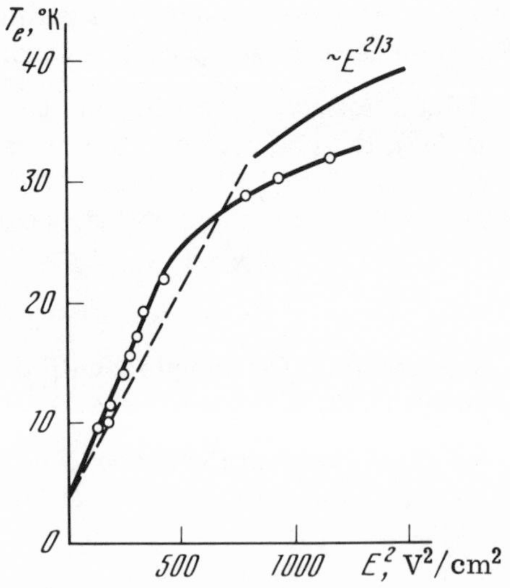

Fig. 44. Dependence of the effective electron temperature on the square of the electric field. The dashed curve represents $T_e = 4.2 + \zeta_{theor} E^2$, where $\zeta_{theor} = 0.034$ deg · cm^2 · V^{-2}; the initial part of the continuous curve represents the values of T_e calculated for $\zeta_{exp} = 0.045$ deg · cm^2 · V^{-2}; circles are the results of a comparison of the experimental data obtained in weak and strong electric fields.

If $\mu \propto AT_e^{1/2}$, which is typical of the momentum relaxation due to the interaction with dipoles, the energy balance condition (5.15) becomes

$$e\mu E^2 = eAT_e^{1/2}E^2 = 2.07 \cdot 10^{-9}\, T_e^{3/2}(1 - T/T_e), \tag{5.16}$$

where the numerical coefficient is in cgs esu or

$$eAE^2 = 2.07 \cdot 10^{-9}\,(T_e - T),$$

and hence

$$T_e = T + \zeta E^2,$$

where

$$\zeta = A\, e/2.07 \cdot 10^{-9}.$$

Bearing in mind that $A = \mu_0/\sqrt{T}$, we find for sample 2 at T = 4.2°K that $\zeta = 3.8 \cdot 10^{-4}\mu_0$ deg · sec · V^{-1} and

$$T_e = 3.8 \cdot 10^{-4}\, \mu_0 \cdot E^2 + T, \tag{5.17}$$

i.e., the relationship is the same as that given by Eqs. (5.13) and (5.11).

Measurements carried out on our samples in a weak electric field (Chap. III, § 2) demonstrate that the scattering by dipoles predominated only at T < 25°K. Therefore, the electron temperature represented by Eqs. (5.11) and (5.17) applies only at these temperatures. At T > 30°K, the scattering by individual charged centers predominates and, in accordance with the data reported in Chap. III, the electron mobility is

$$\mu = \mu_0\,(T_e/30°\,K)^{3/2} = BT_e^{3/2}.$$

We shall consider the dependence of the effective electron temperature on the electric field by deriving an energy balance condition similar to that given by Eq. (5.15). We note that this implies a Maxwellian distribution of the electron velocities with an effective temperature T_e [97], which is generally invalid. However, if the electron density is higher than a certain critical value, the exchange of energy between electrons themselves becomes more effective than the exchange of energy between electrons and the lattice (see also Chap. IV, § 5). In this case a definite electron temperature is established before electrons begin to lose energy. In the case of energy losses to acoustic phonons in gallium arsenide at a lattice temperature of $T = 4.2$°K an estimate gives a critical density $n_{cr} \sim 10^{11}$ cm^{-3} [75, 93]. If the losses are due to optical phonons, then $n_{cr} \sim 10^{10}$ cm^{-3} [94]. Since the electron density in our samples is $n > 10^{15}$ cm^{-3}, we can a priori introduce an effective temperature T_e in a strong electric field.

The energy balance equation for a stronger dependence of the mobility of electrons on their temperature, which applies in the case of scattering by ionized impurities, shows that the energy losses to acoustic phonons do not limit the electron temperature rise:

$$e\mu E^2 = BT_e^{3/2}E^2 = CT_e^{3/2}\left(1 - \frac{T}{T_e}\right),$$

and hence $T_e = CT(C - BE^2)$.

At a certain electric field $E \rightarrow (C/B)^{1/2}$ the electron temperature becomes $T_e \rightarrow \infty$. However, this does not occur in practice because when the losses via acoustic phonons become ineffective in limiting the rise of the electron temperature, the losses via optical phonons become important and – according to Oliver's calculations [102] – they are effective at temperatures $T_e > 30$°K when the momentum is transferred to impurities. The energy losses via optical phonons increase rapidly with rising electron temperature as $\exp(-T_D/T_D)$, where T_D is the Debye temperature [102], and even when this mechanism is relatively weak, it has a considerable effect on the electron temperature. Thus, it is likely that the rise of the electron temperature with increasing field is governed by some combination of these energy loss mechanisms. We shall now try to determine this rise on the basis of the experimental dependence $\mu(E)$ plotted in Fig. 43.

By analogy with the fields corresponding to $T_e < 25$°K we shall assume that in stronger fields the mobility depends again only on T_e. Therefore, a given value of the mobility can be reached either by the application of an electric field or by heating a crystal to $T = T_e$. Hence, we can compare the values of T_e and E corresponding to the same mobility, as demonstrated by circles plotted in Fig. 44. This makes it possible to follow the dependence of T_e on E outside the range of fields used in our measurements.

It is clear from Fig. 44 that an electron temperature $T \approx 25$°K is reached in a field $E \approx 25$ V/cm. In fields $E > 25$ V/cm, the rise of T_e with the electric field slows down, which shows that a new (optical phonon) energy loss mechanism becomes active, in agreement with Oliver's calculations [102]. Direct measurements of the field dependence of the mobility in the range $E > 25$ V/cm show no significant change (see Fig. 43). However, the current–voltage characteristic of sample 1 (curve 1 in Fig. 40), recorded at higher field intensities, shows that in the range $E > 25$–30 V/cm the dependence $j = f(E)$ becomes weaker. This means that, although in the impurity scattering range the mobility rises faster with the electron temperature than in the dipole scattering range, experiments carried out in the range of fields $E > 30$ V/cm show that the mobility may increase more slowly with the field because the rate of rise of T_e with the field E slows down.

The dependence of T_e on E, obtained by comparing the experimental data in weak and strong electric fields (Fig. 44), can be used to determine the contribution of optical phonons to the sum of the energies lost to both types (acoustic and optical) of phonon:

$$e\mu E^2 = \left(\frac{d\varepsilon}{dt}\right)_{ac} + X\left(\frac{d\varepsilon}{dt}\right)_{op}. \tag{5.18}$$

The energy losses per second due to the optical phonons are [102]

$$\left(\frac{d\varepsilon}{dt}\right)_{\text{op}} = eE_0\left(\frac{2kT_D}{m^*}\right)^{1/2} e^{-T_D/T_e},$$

where E_0 is the characteristic electric field given by

$$E_0 = \frac{(\varkappa_\infty^{-1} - \varkappa_0^{-1})\, ekT_D m^*}{\hbar^2}.$$

Here, $\varkappa_0$ and $\varkappa_\infty$ are the static and high-frequency permittivities, respectively. For $\varkappa_0 = 12.53$, $\varkappa_\infty = 10.90$ [94], and $T_D = 345°K$ [87], the characteristic field in GaAs is $E_0 = 15.7$ cgs esu = $4.7 \cdot 10^3$ V/cm. Hence,

$$\left(\frac{d\varepsilon}{dt}\right)_{\text{op}} = e \cdot 6{,}07 \cdot 10^8 e^{-\frac{345°}{T_e}}. \tag{5.19}$$

It follows from Eqs. (5.16), (5.18), and (5.19) that in the range $T_e > 30°K$ the electron temperature depends on the field as shown by circles in Fig. 44, provided $X = 0.05$. Thus, even a very small contribution of the optical phonon scattering (5%) slows down the rise of $T_e = f(E)$ quite considerably. For comparison, Fig. 44 includes the dependence $T_e \propto E^{2/3}$ and we can see that at high values of E this sample does exhibit approximately such a relationship. In this case we have $\mu \propto T_e^{3/2} \propto E$, which follows from the experimental dependence $\mu(E)$ shown in Fig. 43.

For curve 2 in Fig. 40 the dependence j(E) in the range $E > 25$ V/cm becomes stronger than $j \propto E^2$, which corresponds to a faster rise of $\mu(E)$ than $\mu \propto E$. This is clearly due to an even smaller contribution of the optical-phonon-induced losses when electrons are heated to temperatures $T_e > 30°K$.

The Boltzmann kinetic (transport) equation is assumed in our calculations dealing with the explanation of the experimental results. This is not fully justified because under our experimental conditions the mobilities are relatively low, of the order of 100 $cm^2 \cdot V^{-1} \cdot sec^{-1}$, i.e., the relaxation times are short and, therefore, the condition of validity of this kinetic equation may not be satisfied. However, the use of the perturbation theory in combination with this kinetic equation is still the only approach to transport processes in strong electric fields [105].

Calculations based on this approach give results which are in good agreement with the experimental data. Since the calculation of T_e in accordance with Eqs. (5.11) and (5.17) does not involve any arbitrary coefficients but only quantities which are well known for gallium arsenide, it follows that the quantitative agreement between the calculations and experimental results can hardly be regarded as accidental. Clearly, the description of the observed phenomena with the aid of the Boltzmann kinetic equation is a good approximation if it is assumed that electrons of temperatures $T_e < 30°K$ are scattered by dipoles and they lose their energy because of interaction with phonons to which only the deformation potential makes a contribution.

The agreement between the calculated and experimental results shows that current-carrying electrons in our materials are the conduction-band electrons with a mass equal to the mass of electrons in the undistorted conduction band of gallium arsenide: $m^* = 0.07m_0$. This means that in a wide range of energies the energy dependence of the density of states in strongly compensated GaAs is identical with the corresponding dependence applicable to pure GaAs and that the distortions of the spectrum described in Chap. III, §§ 4, 5 extend over only a narrow energy range (~5°K).

CONCLUSIONS

1. An investigation was made of the electrical conductivity and Hall effect in compensated gallium arsenide samples with approximately the same donor and acceptor concentrations but different degrees of compensation of impurities right up to 99%. All the experimental results obtained in the temperature range $0.5 \leqslant T \leqslant 300°K$ could be explained on the model of electrical conduction due to free electrons. In the range $10 \leqslant T \leqslant 300°K$ the electron density was found to be constant. Cooling from ~10 to ~1.5°K reduced the electron density, which was attributed to the localization of some of the free electrons. In the range $0.5 \leqslant T \leqslant 1.5°K$ the electron density ceased to depend on temperature because of the degeneracy of the free-electron gas.

The mobility of free electrons was governed by scattering by ionized impurities in the range $20 \leqslant T \leqslant 100°K$: $\mu \propto T^{3/2}$. In the range $20 \geqslant T \geqslant 1.5°K$ it was governed by the scattering on groups of impurities (dipoles): $\mu \propto T^{1/2}$, and below 1.5°K there was a tendency for the mobility to remain independent of temperature.

2. The experimental data on the temperature dependence of the free-electron density were used to construct an energy scheme in which a band of localized states adjoined the bottom of the free-electron (conduction) band. In some specific cases calculations were made of the density of states in the free-electron and localization bands; the width of the localized-states band, Fermi level position, etc. were also found.

3. A phenomenon analogous to the Mott transition, occurring at an electron density of about $5 \cdot 10^{16}$ cm^{-3} (which was in agreement with the theoretical value), was exhibited by the investigated samples.

4. A model was considered in which it was assumed that electrons were localized in potential wells formed by fluctuations in the donor and acceptor distributions. The number of potential wells capable of localization was calculated on the basis of this model and found to be in agreement with the experimentally determined number of localized electrons.

5. An "induced Mott transition" was observed when samples with partly localized electrons were illuminated with ruby laser radiation. The transition was mainfested as a jump of the photocurrent on increase of the illumination intensity to a certain value. The photocarriers and the equilibrium electrons participating in the current were then joined by delocalized electrons. Quantitative estimates indicated that this jump occurred when the total electron density was close to that predicted by the Mott criterion.

6. It was established that heating of the conduction electrons began in fields $E \geq 2$ V/cm. The dependences of the effective temperature of electrons and of their mobility on the electric field were determined. The mass of hot electrons was found to be close to the effective mass of free electrons in gallium arsenide, in agreement with the proposed free-electron conduction model.

7. When electrons were heated to an effective temperature $T_e < 30°K$ in a crystal whose lattice temperature was 4.2°K, the electron energy losses were mainly due to the scattering by the deformation potential.

LITERATURE CITED

1. M. N. Alexander and D. F. Holcomb, Rev. Mod. Phys., 40:815 (1968).
2. N. F. Mott and Z. Zinamon, Rep. Prog. Phys., 33:881 (1970).
3. M. H. Cohen, Proc. Tenth Intern. Conf. on Physics of Semiconductors, Cambridge, Mass.,

1970, US Atomic Energy Commission, Washington, D. C. (1970), p. 645.
4. N. F. Mott, Adv. Phys., 16:49 (1967).
5. V. L. Bonch-Bruevich, Nachr. Akad. Wiss. Göttingen Math.-Phys. Kl. 2, No. 9, 195 (1971); V. L. Bonch-Bruevich and A. G. Mironov, Introduction to the Electron Theory of Disordered Semiconductors [in Russian], Moscow State University (1972).
6. B. I. Shklovskii, Fiz. Tekh. Poluprovodn., 7:112 (1973).
7. C. Hilsum and A. C. Rose-Innes, Semiconducting III-V Compounds, Pergamon Press, Oxford (1961).
8. A. Miller and E. Abrahams, Phys. Rev., 120:745 (1960).
9. N. F. Mott and W. D. Twose, Adv. Phys., 10:107 (1961).
10. W. Baltensperger, Philos. Mag., 44:1355 (1953).
11. E. M. Conwell, Phys. Rev., 103:51 (1956).
12. N. F. Mott, Philos. Mag., 6:287 (1961).
13. N. F. Mott, Proc. Phys. Soc. London, Ser. A, 62:416 (1949).
14. N. F. Mott, Contemp. Phys., 14:401 (1973).
15. Proc. Intern. Conf. on the Metal–Nonmetal Transition, San Francisco, 1968, Session V – The Hubbard Hamiltonian, in: Rev. Mod. Phys., 40:790-810 (1968); D. I. Khomskii, Fiz. Met. Metalloved., 29:31 (1970); M. Cyrot, Philos. Mag., 25:1031 (1972).
16. J. Hubbard, Proc. R. Soc., Ser. A, 276:238 (1963); 281:401 (1964).
17. D. Adler, Solid State Phys., 21:1 (1968); D. Adler and H. Brooks, Comments Solid State Phys., 1:145 (1968).
18. N. F. Mott, Rev. Mod. Phys., 40:677 (1968).
19. J. Basinski and R. Olivier, Can. J. Phys., 45:119 (1967).
20. O. V. Emel'yanenko, T. S. Lagunova, D. N. Nasledov, and G. N. Talalakin, Fiz. Tverd. Tela (Leningrad), 7:1315 (1965).
21. E. A. Davis, Solid-State Electron., 9:605 (1966); H. Fritzsche, J. Phys. Chem. Solids, 6:69 (1958).
22. N. F. Mott, Adv. Phys., 16:49 (1967).
23. N. F. Mott, J. Non-Cryst. Solids, 8-10:1 (1972).
24. P. W. Anderson, Phys. Rev., 109:1492 (1958).
25. N. F. Mott, Philos. Mag., 17:1259 (1968).
26. N. F. Mott, Philos. Mag., 22:7 (1970).
27. D. J. Thouless, J. Non-Cryst. Solids, 8-10:461 (1972).
28. D. J. Thouless, J. Phys. C, 3:1559 (1970); E. N. Economou and M. H. Cohen, Mater. Res. Bull., 5:577 (1970); J. T. Edwards and D. J. Thouless, J. Phys. C, 5:807 (1972); R. Abou-Chacra, D. J. Thouless, and P. W. Anderson, J. Phys. C, 6:1734 (1973).
29. N. F. Mott, Philos. Mag., 26:1015 (1972).
30. M. H. Cohen, J. Non-Cryst. Solids, 4:391 (1970); Phys. Today, 24(5):26 (1971); T. P. Eggarter and M. H. Cohen, Phys. Rev. Lett., 27:129 (1971).
31. E. A. Davis and W. D. Compton, Phys. Rev., 140:A2183 (1965).
32. N. F. Mott and E. A. Davis, Philos. Mag., 17:1269 (1968)
33. B. I. Shklovskiĭ and A. L. Éfros, Zh. Eksp. Teor. Fiz., 60:867 (1971).
34. B. I. Shklovskiĭ and A. L. Éfros, Zh. Eksp. Teor. Fiz., 61:816 (1971).
35. B. I. Shklovskiĭ, Fiz. Tekh. Poluprovodn., 6:1197 (1972).
36. N. F. Mott, Philos. Mag., 19:835 (1969).
37. F. R. Allen and C. J. Adkins, Philos. Mag., 26:1027 (1972).
38. A. R. Gadzhiev, S. M. Ryvkin, and I. S. Shlimak, Pis'ma Zh. Eksp. Teor. Fiz., 15:605 (1972).
39. I. S. Shlimak and E. I. Nikulin, Pis'ma Zh. Eksp. Teor. Fiz., 15:30 (1972).
40. A. G. Zabrodskiĭ, A. N. Ionov, R. L. Korchazhkina, and I. S. Shlimak, Fiz. Tekh. Poluprovodn., 7:1914 (1973).

41. E. M. Gershenzon, V. A. Il'in, I. N. Kurilenko, and L. B. Litvak-Gorskaya, Fiz. Tekh. Poluprovodn., 6:1687 (1972).
42. E. Wigner, Trans. Faraday Soc., 34:678 (1938).
43. L. L. Foldy, Phys. Rev. B, 3:3472 (1971).
44. P. Fazekas, Solid State Commun., 10:175 (1972).
45. D. Pines, Elementary Excitations in Solids, Benjamin, New York (1963).
46. W. J. Carr, Jr., Phys. Rev., 122:1437 (1961).
47. N. H. March, S. Sampanthar, and W. H. Young, The Many-Body Problem in Quantum Mechanics, Cambridge University Press (1967).
48. W. J. Carr, Jr., and A. A. Maradudin, Phys. Rev., 133:A371 (1964).
49. P. Noziéres and D. Pines, Phys. Rev., 111:442 (1958).
50. J. Durkan, N. H. March, and R. J. Elliott, Rev. Mod. Phys., 40:812 (1968).
51. C. M. Care and N. H. March, J. Phys. C, 4:L372 (1971).
52. E. H. Putley, Proc. Phys. Soc. London, 76:802 (1960); R. K. Willardson and A. C. Beer (eds.), Semiconductors and Semimetals, Vol. I, Academic Press, New York (1966).
53. Y. Yafet, R. W. Keyes, and E. N. Adams, J. Phys. Chem. Solids, 1:137 (1956).
54. D. J. Somerford, J. Phys. C, 4:1570 (1971).
55. E. O. Kane, Phys. Rev., 131:79 (1963).
56. L. V. Keldysh, Doctoral Thesis, Moscow (1965).
57. T. N. Morgan, Phys. Rev., 139:A343 (1965).
58. G. Lucovsky, Solid State Commun., 3:105 (1965).
59. B. I. Shklovskii and A. L. Éfros, Fiz. Tekh. Poluprovodn., 4:305 (1970).
60. L. V. Keldysh and G. P. Proshko, Fiz. Tverd. Tela (Leningrad), 5:3378 (1963).
61. L. M. Falicov and M. Cuevas, Phys. Rev., 164:1025 (1967).
62. N. V. Zavaritskii, Prib. Tekh. Eksp., No. 2, 140 (1956).
63. A. C. Rose-Innes, Low Temperature Techniques: The Use of Liquid Helium in the Laboratory, English Universities Press, London (1964).
64. E. C. Kerr and R. D. Taylor, Ann. Phys. (N. Y.), 20:450 (1962).
65. A. B. Fradkov, Handbook on Physicotechnical Basis of Cryogenics [in Russian], Énergiya, Moscow (1973).
66. E. A. Bobrova, Thesis for Candidate's Degree, Moscow (1971).
67. R. K. Willardson and A. C. Beer (eds.), Semiconductors and Semimetals, Vol. 3, Academic Press, New York (1967).
68. E. A. Movchan and N. N. Bondar', Prib. Tekh. Eksp., No. 3, 226 (1965).
69. B. M. Vul, É. I. Zavaritskaya, I. D. Voronova, and N. V. Rozhdestvenskaya, Fiz. Tekh. Poluprovodn., 5:943 (1971).
70. B. M. Vul, B. L. Voronov, I. D. Voronova, É. I. Zavaritskaya, and N. V. Rozhdestvenskaya, Fiz. Tekh. Poluprovodn., 8:1507 (1974).
71. H. Ehrenreich, Phys. Rev., 120:1951 (1960).
72. F. J. Blatt, Solid State Phys., 4:200 (1957).
73. R. S. Crandall, Solid State Commun., 7:1575 (1969).
74. J. R. Sandercock, Solid State Commun., 7:721 (1969).
75. D. J. Oliver, Phys. Rev., 127:1045 (1962).
76. E. M. Conwell and V. F. Weisskopf, Phys. Rev., 77:388 (1950).
77. H. Brooks, Phys. Rev., 83:879 (1951); C. Herring (unpublished).
78. R. Stratton, J. Phys. Chem. Solids, 23:1011 (1962).
79. A. G. Samoilovich and M. V. Nitsovich, Fiz. Tverd. Tela (Leningrad), 5:2981 (1963); A. A. Tsertsvadze, Fiz. Tekh. Poluprovodn., 3:409 (1969).
80. L. D. Landau and E. M. Lifshitz, Quantum Mechanics: Non-Relativistic Theory, 2nd ed., Pergamon Press, Oxford (1965).

81. E. Jahnke, F. Emde, and F. Lösch, Tables of Higher Functions, McGraw-Hill, New York (1960).
82. B. M. Vul, É. I. Zavaritskaya, I. D. Voronova, G. N. Galkin, and N. V. Rozhdestvenskaya, Fiz. Tekh. Poluprovodn., 7:1942 (1973).
83. R. J. von Gutfeld, "Heat pulse transmission," in: Physical Acoustics: Principles and Methods (ed by W. P. Mason), Academic Press, New York (1968), p. 233.
84. V. Narayanamurti and C. M. Varma, Phys. Rev. Lett., 25:1105 (1970).
85. M. G. Holland, Phys. Rev., 134:A471 (1964).
86. G. J. Lasher and W. V. Smith, IBM J. Res. Dev., 8:532 (1964).
87. J. C. Holste, Phys. Rev. B, 6:2495 (1972).
88. T. C. Cetas, C. R. Tilford, and C. A. Swenson, Phys. Rev., 174:835 (1968).
89. J. Wilks, The Properties of Liquid and Solid Helium, Clarendon Press, Oxford (1967).
90. N. S. Snyder, Cryogenics, 10:89 (1970).
91. M. N. Gurnee, M. Glicksman, and P. Won Yu, Solid State Commun., 11:11 (1972).
92. G. N. Mikhailova, Thesis for Candidate's Degree, Moscow (1974).
93. H. Fröhlich and B. V. Paranjape, Proc. Phys. Soc. London, Sect. B, 69:21 (1956).
94. R. Stratton, Proc. R. Soc., Ser. A, 246:406 (1958).
95. Yu. M. Popov, Doctoral Thesis, Moscow (1963).
96. O. N. Krokhin and Yu. M. Popov, Zh. Eksp. Teor. Fiz., 38:1589 (1960).
97. O. Madelung, Physics of III-V Compounds, Wiley, New York (1964).
98. J. Bardeen and W. Shockley, Phys. Rev., 80:72 (1950).
99. V. L. Ginzburg and A. V. Gurevich, Usp. Fiz. Nauk, 70:201 (1960).
100. B. M. Vul, É. I. Zavaritskaya, I. D. Voronova, and N. V. Rozhdestvenskaya, Fiz. Tekh. Poluprovodn., 7:1766 (1973).
101. E. M. Conwell, High Field Transport in Semiconductors, Suppl. 9 to Solid State Phys., Academic Press, New York (1967).
102. D. J. Oliver, Proc. Sixth Intern. Conf. on Physics of Semiconductors, Exeter, England, 1962, publ. by The Institute of Physics, London (1962), p. 133.
103. A. Zylbersztejn, Proc. Seventh Intern. Conf. on Physics of Semiconductors, Paris, 1964, Vol. 1, Physics of Semiconductors, publ. by Dunod, Paris; Academic Press, New York (1964), p. 505.
104. V. A. Chuenkov, Fiz. Tverd. Tela (Leningrad), 2:799 (1960).
105. J. M. Ziman, Electrons and Phonons, Clarendon Press, Oxford (1960).

INVESTIGATION OF SPONTANEOUS AND COHERENT RADIATION EMITTED FROM INDIUM ANTIMONIDE AS A RESULT OF DOUBLE INJECTION OF CARRIERS

S. P. Grishechkina

A study was made of the spontaneous and coherent radiation emitted from pure and doped indium antimonide due to double injection of carriers. The relationships obtained indicated that, at high injection rates, light holes made a considerable contribution to the electric conductivity and recombination radiation emitted from pure crystals. The application of a magnetic field resulted in a redistribution of carriers between the light- and heavy-hole subbands, which lowered considerably the threshold current for the generation of coherent radiation. Doped crystals exhibited a dependence of the forbidden bandwidth on the acceptor concentration, due to the Coulomb interaction of free carriers with charged impurity centers. Coherent radiation was emitted at 20°K and it was found that the frequency of this radiation could be tuned by the application of a magnetic field and also by altering the dopant concentration.

INTRODUCTION

Development of high-power semiconductor sources of spontaneous and coherent infrared radiation is one of the most important current tasks in semiconductor technology. Semiconductor lasers based on carrier injection are among the most important sources of coherent radiation. Injection lasers have the advantage over other lasers because electric energy is converted directly into radiation and this ensures a high efficiency. However, technical applications of the injection lasers are limited by their relatively low output power.

This output power can be increased by lowering the threshold current, reducing the laser beam divergence, ensuring single-mode emission, etc. In spite of the fact that the first GaAs lasers were built in 1962 [1, 2] and the first InSb lasers appeared in 1964-1965 [3-5], many of these ways of increasing the output power are still in the research stage.

The first sources of coherent radiation made of pure InSb crystals demonstrated that the use of pure materials was one of the ways of reducing the laser beam divergence. In semiconductor lasers this reduction in the divergence angle δ requires an increase in the size of the stimulated emission region d, because beam divergence is governed mainly by diffraction. The divergence angle is $\theta \approx \lambda / d$, where λ is the emission wavelength.

At the time when the present investigation was started, coherent radiation from InSb was obtainable only at temperatures $T \lessdot 10°K$ in strong magnetic fields. The reported threshold current densities exceeded 10^4 A/cm^2.

The present author set herself a task of reducing the threshold current and increasing the operating temperature of the InSb lasers, and finding the optimal conditions for the injection of nonequilibrium carriers into $p^+ - p - n^+$ structures.

This required investigation of the spontaneous and coherent emission from pure and doped InSb crystals and of the electrical conductivity of p^+-p-n^+ structures, and also of the influence of external factors (electric and magnetic fields) on the conductivity and recombination radiation.

Most of the studies of the electrical conductivity and radiative recombination in InSb were carried out under near-equilibrium conditions whereas generation of coherent radiation in pure indium antimonide crystals has been achieved only at injection levels such that the injected-carrier densities have been much higher than the intrinsic density. Thus, investigations of the electrical conductivity and radiative recombination at high injection levels in pure indium antimonide crystals are of basic importance because many parameters of the electron–hole plasma at T = 4.2°K are not yet known. For example, the diffusion length of carriers at high injection levels, their lifetime, radiative transition mechanisms, etc. are still to be determined.

CHAPTER I

REVIEW OF THE LITERATURE

§ 1. Energy Band Structure of Indium Antimonide

Indium antimonide is a semiconductor with the zinc-blende structure. Calculations of the energy bands of crystals with the zinc-blende crystal structure were made by Dresselhaus [6] and Parmenter [7]. According to their calculations, the energy band structure of these crystals is characterized by the following distinguishing features. Spin degeneracy should be lifted at all nonspecific points in the valence Brillouin zone. The top of the valence band is not an extremum and if the subband Γ_{15} is located above all the other valence subbands, the valence band maxima are along the ⟨111⟩ direction. According to Dresselhaus, the energy surfaces near the maxima are ellipsoids of revolution about the ⟨111⟩ axes.

In 1957, Kane [8] proposed a band structure model of indium antimonide. Kane based his model on the assumption that the minimum of the forbidden band at **k** = 0 is absolute and the considerable curvature of the bottom of the conduction band at **k** = 0 is due to the interaction with the valence band at **k** = 0. Hence, he considered only eight wave functions: two S functions with different spins and six P functions. In the first-order approximation of the perturbation theory [9], Kane obtained four energy eigenvalues at **k** = 0. These eigenvalues represented the conduction band, the light- and heavy-hole valence bands, which are degenerate at **k** = 0, and the spin–orbit-split valence band separated by a gap Δ. All these bands are twofold-degenerate in spin. Allowance, in the second order of the perturbation theory, for the interactions with higher bands lifts the valence band degeneracy at **k** = 0 and the splitting is proportional to **k**. Allowance for higher and lower bands lifts the spin degeneracy of the conduction band and of the light-hole valence band along all directions, except ⟨111⟩ and ⟨100⟩, and also lifts the spin degeneracy of the heavy-hole valence band everywhere, even at **k** = 0, with the exception of the ⟨100⟩ direction. For the simplest direction ⟨110⟩ the dispersion laws of the indium antimonide energy bands can be expressed as follows:

$$\mathcal{E}_c = \mathcal{E}_g + \frac{\hbar^2 k^2}{2m_0}\left(1 + \frac{m_0}{m}\right), \quad (1)$$

$$\mathcal{E}_{v_1} = -\frac{\hbar^2 k^2}{2m_0}\left(\frac{m_0}{m} - 1\right), \quad (2)$$

$$\mathcal{E}_{v_2} = -\frac{\hbar^2 k^2}{2m_0}\left(\frac{m_0}{m}\frac{2(\mathcal{E}_g + \Delta)}{3\mathcal{E}_g + 2\Delta} - 1\right), \quad (3)$$

$$\mathcal{E}_{v_3} = -\Delta - \frac{\hbar^2 k^2}{2m_0}\left(\frac{m_0}{m}\frac{\mathcal{E}_g}{3\mathcal{E}_g + 2\Delta} - 1\right). \quad (4)$$

Here, $\mathcal{E}_c$, $\mathcal{E}_{v_1}$, $\mathcal{E}_{v_2}$, and $\mathcal{E}_{v_3}$ are the energies in the conduction band, heavy- and light-hole valence bands, and spin–orbit-split band; m_0 is the mass of a free electron; m is a calculation parameter. At the bottom of the conduction band, where this band is parabolic, the effective electron mass is given by

$$1/m_e = 1/m_0 + 1/m. \tag{5}$$

Figure 1 shows the dependences of the energies (averaged over the directions) in the conduction and valence bands on $\mathbf{k}^2$, deduced for InSb in accordance with the Kane theory.

The results of these calculations can be summarized as follows.

1. The conduction band of indium antimonide is spherically symmetric, strongly non-parabolic, and spin-degenerate at $\mathbf{k} = 0$. The effective mass at the bottom of the conduction band is approximately $m_e \approx 0.013m_0$. The influence of the higher bands lifts the spin degeneracy because of the terms proportional to $\mathbf{k}^3$ and this is true of all directions except ⟨111⟩ and ⟨100⟩. A small coefficient in front of the terms with $\sim\mathbf{k}^3$ lifts the spin degeneracy only for $\mathscr{E} > 0.1$ eV.

2. The maximum of the heavy-hole valence band v_1 is shifted along the ⟨111⟩ direction by 0.3% of the distance from the boundary of the Brillouin zone and by $\sim 10^{-4}$ eV relative to $\mathbf{k} = 0$. This valence band is not affected by the conduction band. The effective mass is governed by the influence of the remaining bands. A strong influence of these bands lifts the spin degeneracy even at $\mathbf{k} = 0$ because of terms linear in $\mathbf{k}$ and this is true everywhere with the exception of the ⟨100⟩ direction. This band is strongly anisotropic and the mass ratio is $m_{\parallel} : m_{\perp} = 3$. The average density-of-state effective mass of the heavy holes is estimated to be $0.011m_0$.

3. The light-hole valence band v_2 has its maximum at $\mathbf{k} = 0$. Like the conduction band, it is spherically symmetric but it is not parabolic. The interaction with the other bands lifts the degeneracy because of terms linear in $\mathbf{k}$ everywhere except along the ⟨111⟩ and ⟨100⟩ directions. The effective mass in the light-hole band at $\mathbf{k} = 0$ is estimated by Kane to be $0.011m_0$.

4. The band v_3 split by the spin–orbit interaction is in almost all respects analogous to the v_2 band and it is located 0.9 eV below the v_2 band. The only difference between the two bands is the absence of terms linear in $\mathbf{k}$, i.e., there is no lifting of the spin degeneracy. The effective mass at $\mathbf{k} = 0$ is $m_3 \approx 0.012m_0$. Figures 2 and 3 show the energy scheme of the valence band and constant-energy section for the light- and heavy-hole bands taken from [10].

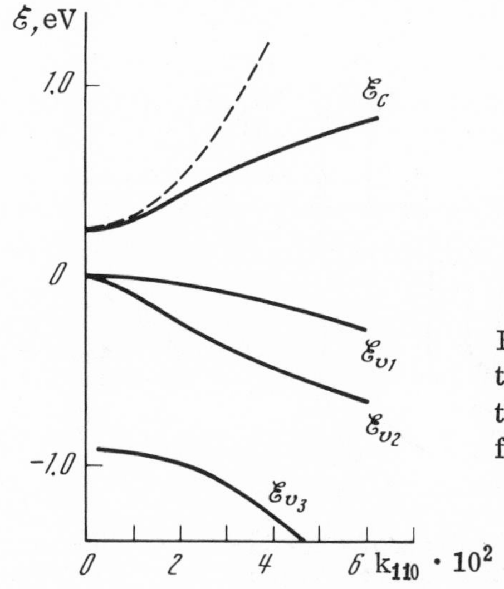

Fig. 1. Dependence of the carrier energy on the wave number along the ⟨110⟩ direction in the conduction and valence bands in InSb (taken from [8]).

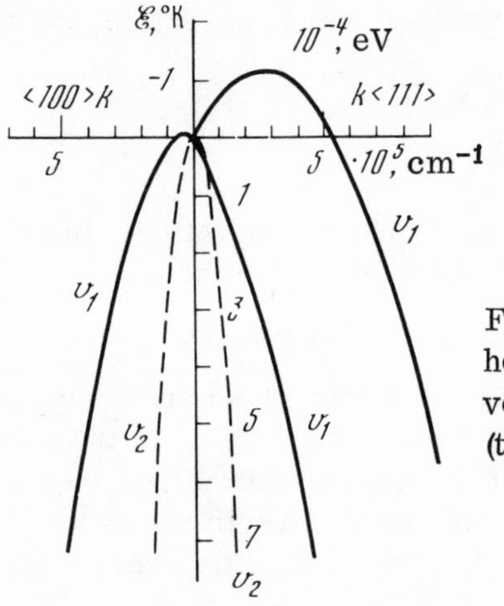

Fig. 2. Dependences of the light- and heavy-hole energies on the reciprocal of the wave vector along the ⟨100⟩ and ⟨111⟩ directions (taken from Ohmura's work [10]).

§ 2. Absorption, Magnetoabsorption, and Recombination Radiation of Indium Antimonide Near the Fundamental Absorption Edge

The probability of a transition of a carrier from one state to another as a result of some perturbation is described by

$$W_i = \frac{2\pi}{\hbar} |M_{ij}|^2 \rho(\mathscr{E}_f). \tag{6}$$

Here, $|M_{ij}|$ is the matrix element of the perturbation linking the initial and final states; $\rho(\mathscr{E}_f)$ is the density of the final states. When the perturbation is an electromagnetic wave, Eq. (6) includes the optical matrix element M_{ij}. According to the Kane model, the fundamental absorption near the edge of InSb [8, 11] is due to direct transitions between the conduction band, on the one hand, and the light- and heavy-hole valence bands, on the other. Both types of transition are allowed at $\mathbf{k} = 0$ and are governed by the same matrix element of the optical transition $|M_{c,v}| = \text{const}$. This matrix element decreases with increasing wave vector $\mathbf{k}$, but a compari-

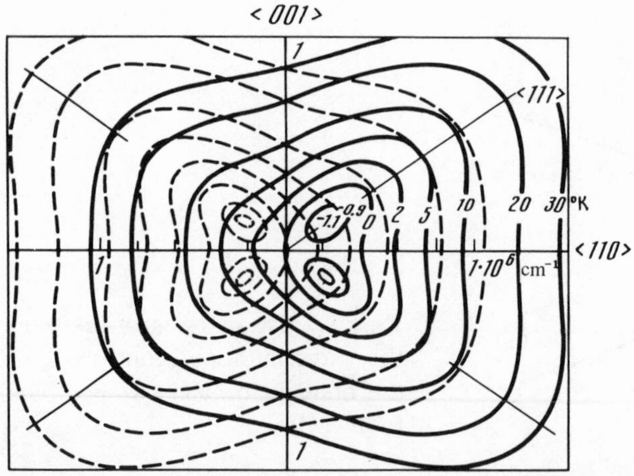

Fig. 3. Section through the constant-energy surface of the heavy holes in indium antimonide [10].

son of the theory with experiment [11] shows that in the case of the electron–heavy hole transitions this has to be allowed for only in the range $\hbar\omega - \mathscr{E}_g > 0.1$ eV (here, $\hbar\omega = h\nu$ is the photon energy; $\mathscr{E}_g$ is the forbidden band width).

Thus, in investigations at the fundamental absorption edge we can assume that the matrix element of a transition is constant and we can calculate the transition probability if we know the density of states $\rho(\mathscr{E}_f)$. The energy $\mathscr{E}_f$, which occurs in Eq.(6), is governed by the difference between the energies of the initial and final states. Since near $\mathbf{k} = 0$ the dispersion laws of all the bands can be regarded as quadratic, the electron energy can be expressed in the form $\mathscr{E}_e = \mathscr{E}_c + \frac{\hbar^2\mathbf{k}^2}{2m_e}$ ($\mathscr{E}_c$ is the energy corresponding to the ottom of the conduction band and m_e is the effective mass of electrons) and the energy of electrons can be described by $\mathscr{E}_p = \mathscr{E}_v - \frac{\hbar^2\mathbf{k}^2}{2m_p}$ ($\mathscr{E}_v$ is the energy of the top of the valence band and m_p is the effective mass of holes). Then, the photon energy is described by

$$h\nu = \mathscr{E}_e - \mathscr{E}_p = \mathscr{E}_g + \frac{\hbar^2\mathbf{k}^2}{2}\left(\frac{1}{m_e} + \frac{1}{m_p}\right) = \mathscr{E}_g + \frac{\hbar^2\mathbf{k}^2}{2m_r}, \tag{7}$$

where

$$m_r = \frac{m_e m_p}{m_e + m_p}. \tag{8}$$

The density of the states participating in the optical transitions is also governed by the reduced effective mass m_r:

$$\rho(\mathscr{E}_f) = \rho(h\nu) = \frac{(2m_r)^{3/2}}{2\pi^2\hbar^3}(h\nu - \mathscr{E}_g)^{1/2}. \tag{9}$$

The optical transition probability is

$$W_i \equiv \alpha h\nu = \frac{2\pi}{h}\,|M_{cv}(\mathbf{k})|^2\,\frac{(2m_r)^{3/2}}{2\pi^2\hbar^3}(h\nu - \mathscr{E}_g)^{1/2}. \tag{10}$$

The absorption coefficient can be deduced from the above formula only if the valence band is completely filled and the conduction band is quite empty. At nonzero temperatures or in the presence of some impurities N_i we cannot use the above simple expression for $\rho(\mathscr{E})$. In such cases the density of states participating in the direct transitions is

$$\rho(\mathbf{k}) = \rho_0(\mathbf{k})\,[f_v(\mathscr{E}_v - \xi) - f_c(\mathscr{E}_c - \xi)], \tag{11}$$

where $\rho_0(\mathbf{k})$ is the density of the final states without allowance for their carrier occupancy; ξ is the Fermi level; $f_v(\mathscr{E}_v - \xi)$ and $f_c(\mathscr{E}_c - \xi)$ are the probabilities that the states in question are occupied. The absorption coefficient of strongly degenerate n-type semiconductors is

$$\alpha(\mathbf{k}) = \alpha_0(\mathbf{k})(f_v - f_c) = \alpha_0(\mathbf{k})\left[1 - \frac{1}{1 + \exp\left(\frac{\mathscr{E}_c - \xi}{kT}\right)}\right]. \tag{12}$$

If the valence band is strongly degenerate, we have to calculate $\alpha(\mathbf{k})$ for each band separately and add the results.

A magnetic field alters basically the densities of states in the bands creating singularities at energies governed by the condition $\mathscr{E}_e = \mathscr{E}_g + (n + 1/2)\hbar\omega_c$, where $\hbar\omega_c = \hbar(eH/m_e c)$ (**H** is the magnetic field; m_e is the effective electron mass; c is the velocity of light). The energy of an electron or a hole in a magnetic field is described by two quantum numbers n and $\mathbf{k}_z$, and it can be expressed in the form

$$\mathscr{E}_e(\mathbf{k}_z, n) = \mathscr{E}_c + \frac{\hbar^2 \mathbf{k}^2}{2m_e} + \hbar\omega_e\left(n + \frac{1}{2}\right), \tag{13}$$

$$\mathscr{E}_p(\mathbf{k}_z, n) = \mathscr{E}_v + \frac{\hbar^2 \mathbf{k}^2}{2m_p} + \hbar\omega_p\left(n + \frac{1}{2}\right). \tag{14}$$

The density of states is defined by $N(\mathscr{E}, \mathbf{H}) = \text{const}\cdot\hbar\omega \sum_{n=0}^{\infty}\left[\mathscr{E} - \left(n + \frac{1}{2}\right)\hbar\omega\right]^{1/2}$. Therefore, in the case of direct allowed transitions the frequency dependence of the absorption coefficient is

$$\alpha(\omega, \mathbf{H}) \propto \hbar\omega_r \sum_n \left(\hbar\omega - \mathscr{E}_g - \left(n + \frac{1}{2}\right)\hbar\omega_r\right]^{1/2}, \tag{15}$$

where

$$\omega_r = \omega_n + \omega_p = eH/cm_r.$$

Allowance for the spin splitting in the radicand in the above expression gives rise to additional terms

$$\pm\, 1/2 g_{n,p}\mu_\beta \mathbf{H}, \tag{16}$$

where $\mu_\beta = {}^1/_2(eh/m_0c)$ is the Bohr magneton and m_0 is the free-electron mass.

Much work has been done on the absorption near the fundamental edge. Gobeli and Fan [12] reported, in 1960, measurements of the absorption in n- and p-type InSb. They established that the electron–light hole transitions could be observed only in relatively pure crystals. In heavily doped samples the free-carrier interband transitions in the valence band produced a background which masked the electron–light hole transitions. Moreover, Gobeli and Fan observed a strong broadening of the absorption edge for the heavy hole–electron transitions. Assuming that the spreading of the threshold for these transitions was due to the corrugation of the heavy-hole valence band and using the dispersion law for this band subject to allowance for the corrugation,

$$\mathscr{E}_{v_1} = \frac{\hbar^2\mathbf{k}^2}{2m_0} - \frac{\hbar^2\mathbf{k}^2}{2m_v}\left(1 - 3\gamma'\,\frac{\mathbf{k}_x^2\mathbf{k}_y^2 + \mathbf{k}_y^2\mathbf{k}_z^2 + \mathbf{k}_z^2\mathbf{k}_x^2}{k^4}\right), \tag{17}$$

Gobeli and Fan determined the nonsphericity parameter γ'. According to their calculations, $\gamma = 3\gamma' = 2$. An analysis of the results led Gobeli and Fan to the conclusion that the corrugation of the heavy-hole valence band was the main cause of the broadening of the absorption edge of n- and p-type samples. Measurements of the properties of the p-type material yielded the effective mass of the light holes on the assumption that the dispersion law was parabolic ($m_2 = 0.12m_0$).

More accurate estimates of the values of the effective masses and the forbidden band width of InSb were deduced from the magnetoabsorption data obtained by Zwerdling et al. [13]. Zwerdling et al. observed oscillations of the absorption coefficient in a magnetic field. They interpreted their results when they allowed not only for the spin splitting, in accordance with Eq. (16), but also for the band nonparabolicity and the Coulomb interaction (excitons). Extrapolation to zero magnetic field gave the forbidden band width of indium antimonide ($\mathscr{E}_g$ = 0.2357 ± 0.0005 eV) and the effective masses of the light and heavy holes m_2 and m_1. Later, Pidgeon and Brown [14] calculated the absorption maxima allowing for the degeneracy of the valence band at **k** = 0 and determined the magnetoabsorption at T = 4.2°K in polarized light in the **E** ∥ **H** and **E** ⊥ **H** configurations. They calculated the relative intensities of the absorption lines for the $a^{\pm}(n)b^{\pm}(n)$ transition series (the notation is due to Dresselhaus [15] and Luttinger [16]). However, Pidgeon and Brown did not allow for terms linear in **k**, which would have resulted in the lifting of the degeneracy at **k** = 0 and of the forbiddenness of some of the transitions [17].

Recombination radiation emitted from indium antimonide was investigated quite thoroughly [18-27], beginning from the early papers of Moss [18] carried out at room temperature and relatively late studies of Phelan [19], who investigated the dependences of the spectral characteristics of the radiation at T = 4.2°K on the concentration and type of impurities in InSb. From our point of view, of greatest interest were the investigations carried out at liquid helium temperature. A study of the recombination radiation spectra at liquid helium temperature revealed electron–acceptor level radiative transitions ($\mathscr{E}_{max} \approx 228$ meV) [20], optical-phonon-assisted transitions ($\mathscr{E}_{max} \approx 212$ meV) [21], as well as transitions of electrons from donor levels to the valence band in a magnetic field [22] and an acoustic-phonon-assisted resonance luminescence line [23]. Coherent radiation was generated in 1965 in strong magnetic fields [5, 24] and this gave an impetus to intensive investigations of the magnetoluminescence of indium antimonide [25-27]. The g factor of electrons was determined more accurately by observing radiative trnansitions of electrons from the $m_s = \pm 1/2$ spin sublevels of the zeroth Landau level to the heavy-hole valence band in magnetic fields H > 10 kOe.

However, none of these investigations of the recombination radiation emitted from indium antimonide [18-27] revealed radiative transitions of electrons from the conduction band to the light-hole valence band.

§ 3. Electrical Conductivity of p-Type InSb

The electrical conductivity of p-type InSb is due to two types of hole: $\sigma = \sigma_1 + \sigma_2 = \sigma_1 (1 + \sigma_1/\sigma_2)$, where $\sigma_1 = e\mu_1 p_1$, and $\sigma_2 = e\mu_2 p_2$. Here, p_1, p_2, μ_1, μ_2, σ_1, and σ_2 are the densities and mobilities of the heavy and light holes and the electrical conductivities due to these holes. The relative contributions of the light and heavy holes to the electrical conductivity can be estimated if the ratio of their densities and mobilities are known. Since $p_1/p_2 = (m_1/m_2)^{3/2}$, it is sufficient to estimate the ratio of the mobilities. At 4.2°K, even in fairly pure InSb crystals, the impurity scattering μ^I predominates at moderate injection rates and the mobility is then $\mu_i^I \approx A/m_i$ in accordance with [28]. Therefore, for the scattering by impurities we have $\sigma_2/\sigma_1 = (m_2/m_1)^{1/2} \ll 1$, because in the case of InSb we have $m_2/m_1 \ll 1$, i.e., the contribution of the light holes to the conductivity can be ignored.

In the scattering by lattice vibrations we have $\mu_i^L \propto B/m_i^{5/2}C_i^2$ for indium antimonide at a certain temperature [29]. In this case we find that $\sigma_2/\sigma_1 = (m_1/m_2)(C_1^2/C_2^2)$. Here, C_2^2 and C_1^2 are the deformation potential constants of the light and heavy holes, respectively. If

$$\frac{C_1^2}{C_2^2} \ll \frac{m_2}{m_1}, \tag{18}$$

the main contribution to the electrical conductivity should be made by the heavy holes but the inequality (18) is unlikely to be satisfied and it is much more probable that $C_1^2/C_2^2 > m_2/m_1$, but in this case we have $\sigma \approx \sigma_2$. Thus, when carriers are scattered by the lattice vibrations, we can no longer ignore the contribution of the light holes.

At very high hole densities the interband scattering of the light by the heavy holes may become important. This reduces the conductivity, as predicted theoretically in [30] and found experimentally in [31].

It must be mentioned that the forbidden band width of indium antimonide is so narrow and the electron mobility is so high that an inversion of the sign of the Hall coefficient is exhibited by the lightly doped p-type samples when the temperature is increased from 4.2°K to the room value, and at T = 300°K the n-type conduction predominates. Since the intrinsic carrier density in indium antimonide at T = 300°K is

$$n_i = 2\left(\frac{2\pi kT}{h^2}\right)^{3/2} (m_e m_p)^{3/4} \exp\left(-\frac{\mathscr{E}_g}{2kT}\right) \gtrsim 2\cdot 10^{16}\ \text{cm}^{-3}, \tag{19}$$

and the mobility ratio is $b = \mu_n/\mu_p \approx 10^2$, an analysis of the temperature dependences of the Hall effect and electrical conductivity must be made using not the two-band model (light and heavy holes) but the three-band model (light and heavy holes, and electrons). This model was used by Schönwald [32] in a theoretical analysis of the experimental results obtained at high temperatures.

The contribution of the light holes to the valence-band conduction is particularly important in studies of the magnetoresistance of p-type indium antimonide. Investigations have shown that the linear relationship between $\Delta\rho/\rho$ and H^2 breaks down for the p-type material well before the saturation predicted by the classical theory [33-35]. These experiments demonstrated directly the existence of the light- and heavy-hole valence subbands. Champness [36] analyzed all the facts on the basis of the two-band model and found that the light- and heavy-hole mobilities in the purest p-type indium antimonide samples differed by a factor of 7.4 at 77°K and the proportion of the light holes participating in electrical conduction was 1.3% of the total density of holes in the valence band. He estimated the mobilities of the heavy and light holes (μ_1 = 8400 and μ_2 = 62,000 $cm^2 \cdot V^{-1} \cdot sec^{-1}$, respectively). He found that the mobilities of the light and heavy holes were limited, even in the purest crystals, by the scattering on ionized impurities. Thus, in all the experiments carried out under thermodynamic equilibrium conditions the scattering by ionized impurities or optical phonons predominated and in these cases the contribution of the light holes to the electrical conductivity was not very large.

Measurements of the Shubnikov–de Haas effect in heavily doped p-type indium antimonide samples kept at 4.2°K were reported in two recent papers [37, 38]. These measurements revealed 50% oscillations of the current, and the oscillation period expressed in terms of the reciprocal magnetic field corresponded to the effective mass of the light holes. The amplitude of the oscillations (50%) could not be associated with the oscillations of the light-hole electrical conductivity because this conductivity represented 10% of the total. Therefore, the oscillation amplitude was explained by assuming that the application of a magnetic field altered drastically the nature of the interband scattering in the valence band. Every time the light-hole Landau level crossed the Fermi level in the valence band [37] the probability of the interband scattering to the light-hole band increased steeply. Moreover, a maximum corresponding to a transition forbidden by the selection rules was observed in [37]. It was reported in [38] that a negative magnetoresistance appeared when the last light-hole Landau level crossed the Fermi level in the valence band.

The same results were obtained also in the absence of degeneracy in the valence band [38] but in this case a negative magnetoresistance was observed in a magnetic field which caused the light-hole Landau level to cross the energy kT.

§ 4. Carrier Lifetimes in InSb

Indium antimonide belongs to a small group of semiconducting materials (for example, Ge, Si, and GaAs) which can be prepared in a very pure form. The narrow forbidden band of InSb makes it possible to estimate the T = 300°K value of the lifetime in an intrinsic material using the Van Roosbroeck–Shockley formula [39]:

$$\tau_i = \frac{n_i}{2R} = \frac{n_i \int\limits_0^{\infty} \frac{n^3 \alpha U^3}{e^4 - 1}}{2 \cdot 1.785 \cdot 10^{22} \left(\frac{T}{300} \right)^4}, \tag{20}$$

where n is the refractive index; α is the absorption coefficient; $U = h\nu / kT$; the application of this formula gives $\tau_i \approx 0.3 \cdot 10^{-6}$ sec. However, measurements of the lifetime carried out at T = 300°K in sufficiently pure InSb crystals have shown that the actual lifetime is at least three times smaller than this estimate. More detailed investigations carried out on n- and p-type indium antimonide [40-43] demonstrated that in the T > 200°K range the temperature dependence of the carrier lifetime is described satisfactorily by the Auger recombination mechanism [44]. At lower temperatures (T = 77°K) there is a discrepancy between the lifetimes in the n- and p-type materials. This was reported by Wertheim [45]. He discovered that in n-type InSb the photoconductivity decay was exponential at all temperatures, whereas in p-type samples at 77°K the decay curves were described by two time constants. He attributed this behavior of the photoconductivity decay in the p-type material to the capture of carriers by traps. In a later paper of Laff and Fan [46] it was shown that the lifetimes of electrons and holes differed strongly in p-type samples at 77°K. Laff and Fan interpreted their results on the basis of a model with two recombination levels. They investigated a large number of n- and p-type samples and they showed that the probabilities of electron and hole trapping had different temperature dependences.

In all the cited investigations the carrier lifetime was measured by the photoconductivity or photoelectromagnetic (photomagnetic) effect methods [41, 47] at $T \gtrsim 77$°K using n- and p-type samples with total impurity concentrations in the range $N_A + N_D \gtrsim 10^{14}$ cm^{-3}. Recent measurements of the carrier diffusion length demonstrated that $\tau = 3 \cdot 10^{-3}$ sec in very pure p-type indium antimonide at T = 100°K [47]. It was concluded in [47] that the lifetime in these pure samples of indium antimonide ($N_A \approx 10^{11}$ cm^{-3}) was limited by the interband radiative recombination mechanism. This was indicated by the temperature dependence of the lifetime. An increase in the impurity concentration caused the lifetime to fall rapidly to $\tau \approx 10^{-7}$ sec in samples with $p = 10^{16}$ cm^{-3}. The dependence of the carrier lifetime on the impurity concentration at temperatures T > 77°K was also reported by Kurnick and Zitter [41].

The relative contributions of the various recombination mechanisms in n-type InSb at different temperatures and for samples with different impurity concentrations were estimated by Benoit à la Guillaume and Fishman [48] from the temperature and concentration dependences of the quantum efficiency of the luminescence. They analyzed the results on the assumption that the quantum efficiency η was governed by the lifetimes representing the radiative τ_R and nonradiative τ_N processes: $\eta = \tau_N / (\tau_R + \tau_N) \approx \tau_N / \tau_R$. Possible radiative mechanisms considered included the radiative interband recombination in the case of the Boltzmann [39] and Fermi [49] statistics, and the nonradiative Shockley–Read [50] and Auger [44] types of recombination. Measurements and estimates indicated that the capture by impurity centers was the low-temperature nonradiative process which limited the quantum efficiency. At higher temperatures, and also in heavily doped samples, the Auger mechanism predominated. Unfortunately, the method employed could not be used to estimate the lifetimes associated with the individual recombination processes.

Other investigators who estimated the carrier lifetime at 4.2°K were Guseinov, Nasledov, Pentsov, and Popov [51] and Shotov and Muminov [52]. Guseinov et al. [51] used the photo-electromagnetic (photomagnetic) effect method, whereas Shotov and Muminov [52] estimated the lifetime by a method suggested by Konnerth and Lanza [53]. In this method it was assumed that the lifetimes were independent of the injected carrier density. The lifetimes were measured under strongly nonequilibrium conditions, when the injected carrier density was much higher than the intrinsic density. Shotov and Muminov obtained the value $\tau \approx 10^{-5}$ sec for the spontaneous interband radiative recombination at 4.2°K. This lifetime was greater than the radiative lifetime in the doped material [50] but shorter than the lifetime in the intrinsic semiconductor at the same temperature [39], which was to be expected for high injection rates [54].

§ 5. Stimulated Emission from Indium Antimonide

In 1958, Basov, Vul, and Popov [55] predicted that it should be possible to obtain stimulated emission of radiation from semiconductors. They showed that an inversion of the carrier distribution ("negative temperature") is required for laser action in semiconductor quantum oscillators and amplifiers. Since a system with a population inversion is in thermodynamic disequilibrium, it is necessary to introduce the concept of quasi-Fermi levels. This concept can be introduced if the energy relaxation time of carriers is much shorter than their lifetime. Then, in a time much shorter than the lifetime an energy level distribution is established in each of the bands, and this distribution is described by the Fermi–Dirac function in which the temperature is understood to be the lattice temperature and the Fermi level is understood to be the quasi-Fermi level of each of the bands:

$$f_{c,v} = \{1 + \exp[(\mathscr{E} - F_{c,v})/kT]\}^{-1}. \tag{21}$$

The population inversion condition implies that $h\nu < F_e - F_p$, where F_e and F_p are the quasi-Fermi levels of electrons and holes.

The population inversion condition for the transitions between the conduction and valence bands is [55-58]

$$F_e - F_p \gtrsim \mathscr{E}_g. \tag{22}$$

For the transitions between donors and the valence band or between electrons from the conduction band and acceptors the condition (22) can be rewritten in the form [59]

$$F_e - F_p \gtrsim \mathscr{E}_g - \mathscr{E}_I, \tag{23}$$

where $\mathscr{E}_I$ is the ionization energy of a donor or acceptor.

The population in a semiconductor can be inverted by optical [60] and electrical [2] injection or by electron bombardment [61]. In the electrical injection of carriers across a p–n junction a negative-temperature state is established if the bias voltage across the junction V is

$$V \gtrsim \frac{\mathscr{E}_c - \mathscr{E}_v}{e} = \frac{\mathscr{E}_g}{e}. \tag{24}$$

Under real conditions the establishment of a negative-temperature state is a necessary but insufficient condition for obtaining a negative absorption coefficient. We must also ensure that the gain exceeds the sum of all the radiation losses. The gain for transitions between two levels can be described by [62]

$$\alpha(h\nu) = \left(\frac{\lambda/n}{8\pi}\right)^2 \frac{[N_i(g_k/g_i) - N_k]}{\tau} g(\nu), \tag{25}$$

where λ is the wavelength in vacuum; n is the refractive index; g_i and g_k are the statistical weights of the upper and lower states; N_i and N_k are numbers of atoms (in 1 cm^3) at the upper and lower states; τ is the spontaneous emission lifetime for transitions between these states; $g(\nu)$ is the normalized line profile

$$\int g(\nu)\,d\nu = 1; \tag{26}$$

ν is the emission frequency ($\nu = c/\lambda$).

It follows from the above expression that the gain increases with decreasing width of the spontaneous emission line and with decreasing radiative recombination time.

Thus, the necessary and sufficient condition for stimulated emission is a population inversion $F_e - F_p > h\nu$ and an excess of the gain over all the radiation losses in a crystal. Coherent radiation is obtained when a system is self-excited, which can be achieved by providing a feedback with the aid of mirrors forming a resonator. The self-excitation condition for a resonator is

$$KB = 1, \tag{27}$$

where K is the gain and B is the feedback parameter. (In our experiments we used a Fabry–Perot resonator [64] and a four-sided resonator [63].) The stimulated emission threshold is

$$R \exp[(-\alpha_t - \beta)L] = 1. \tag{28}$$

Here, R is the reflectivity of the inner walls of the active regions; β represents the bulk losses; α_t is the threshold value of the absorption coefficient $\alpha(h\nu)$ for $N_i/N_R > 1$.

The threshold value of the power is

$$P_t \gtrsim \frac{N_i}{\tau} h\nu V. \tag{29}$$

It follows from Eqs. (25), (28), and (29) that

$$P_t \gtrsim \frac{8\pi V h n^2 \nu \Delta\nu}{\eta c^3}\left(-\frac{\ln R}{L} + \beta\right). \tag{30}$$

It is clear from the above expression that the narrower the spontaneous line, the higher is the quantum efficiency; moreover, the lower the photon energy, the more likely is the attainment of coherent emission.

Dumke [64] analyzed the optical properties of semiconductors and established that the most suitable materials for the generation of coherent radiation were those with simple bands, which include also InSb.

Benoit à la Guillaume and Lavalard [65] were the first to obtain stimulated emission from indium antimonide at 4.2 and 2°K using alloyed p–n junctions made of pure n-type InSb with an initial electron density $n \approx 10^{14}$ cm^{-3}. Minority carriers were injected by current pulses applied in the forward direction. At high current densities ($\approx 2.1 \cdot 10^5$ A/cm^2) and low temperatures (T = 2°K), these authors obtained very weak coherent radiation modes in the long-wavelength part of the spontaneous emission spectrum.

Phelan and Rediker [3] also obtained stimulated emission in a longitudinal magnetic field of 27 kOe in purer samples of n-type indium antimonide than those used in [65]. A magnetic field was applied because it reduced the threshold current needed for the generation of coherent radiation. Bell and Rogers [4] studied the coherent radiation obtained in a magnetic field of H = 31 kOe when the current density was $j = 1.4 \cdot 10^3$ A/cm^2 at T = 1.7°K. In the present study we also observed coherent radiation only in strong magnetic fields. It was found that the application of H = 40 kOe reduced strongly the threshold current but in higher fields the dependence of I_{th} on H approached saturation. Bell and Rogers [4] applied a strong magnetic field at T = 1.9°K to n-type indium antimonide samples with $n = 8 \cdot 10^{12}$ cm^{-3} and they investigated the coherent radiation modes. In every case these coherent modes were observed only in a strong magnetic field at temperatures in the range $T \leqslant 4.2$°K.

The method of double injection of carriers into pure indium antimonide crystals was found to be very promising from the point of view of population inversion because under these conditions the stimulated emission zone was large (≈50-100 μ) and this ensured weak divergence of the laser beam. In one of the first investigations on pure crystals it was found that coherent radiation was emitted parallel to the direction of the current. A sample was placed on a thin copper heat sink in a helium cryostat; the p^+ contact was formed by the diffusion of zinc and the n^+ contact was produced by liquid epitaxy [65]. Coherent radiation was emitted by these samples at T ≈ 10°K when the current density was $j = 6 \cdot 10^4$ A/cm^2 in magnetic fields of $H \geqslant 7$ kOe. The duration of the injection pulses did not exceed 50 nsec. Melngaillis [66] investigated luminescence and coherent emission from similar p^+-p-n^+ structures in InSb. Coherent radiation was observed at right-angles to the direction of the current. The distribution of the intensity of the recombination radiation was determined along the direction of flow of the current. It was found that the injection luminescence was observed throughout the p-type base ≈400 μ long. The angular distribution of the laser radiation was estimated and it was found that the size of the coherent radiation spot was ≈50 μ.

The considerable reduction of the threshold current in a magnetic field attracted the attention of many authors because the effect was observed in longitudinal and transverse fields. Many experiments were carried out and interesting results were obtained but no satisfactory explanation has yet been provided. However, the effect is very considerable and the threshold current can be reduced by almost an order of magnitude, which is very important from the practical point of view.

CHAPTER II

EXPERIMENTAL METHOD

§ 1. Investigated Samples

Nonequilibrium carriers have to be injected in a sample in studies of spontaneous and coherent radiation. This can be done by electron bombardment [61], optical pumping [60], or electrical injection. In the range of wavelengths corresponding to the fundamental absorption edge of InSb there are no sufficiently powerful radiation sources so that a GaAs semiconductor laser emitting photons of 1.5 eV energy is frequently used as the optical pumping source [2]. In this pumping method, as also in the electron bombardment case, the energy of carriers in the bands at the moment t = 0 is very high and this energy is lost as a result of relaxation due to the interaction with lattice vibrations. Since the effective electron mass in indium antimonide is very small, and, consequently, the interaction with acoustic phonons is weak, the method of electrical injection of carriers is important because then the injected carriers have energies close to the thermal value (particularly when the lattice temperature is T = 4.2°K).

Pulse injection across a p−n junction can be used in studies of spontaneous and coherent emission. However, in the case of p−n junctions the injection-carrier density cannot exceed the lowest carrier density in the n- or p-type region. We investigated pure and Zn-doped InSb samples so that it was necessary to establish conditions for injection of large numbers of relatively cold carriers into pure InSb crystals. This could be done by the double injection method via two rectifying contacts in p^+-p-n^+ structures. We investigated samples with carrier densities from $p_0 = N_A - N_D = 2 \cdot 10^{13}$ cm^{-3} to $p_0 = N_A - N_D = 8 \cdot 10^{15}$ cm^{-3}. These structures could be prepared by diffusion or alloying. The diffusion method was convenient because the diffusion front was governed by the quality of the surface finish through which the diffusion took place. Unfortunately, the diffusion method could not be applied to p-type indium antimonide because Te deposited on the surface of this semiconductor did not diffuse as a result of heating but formed a glassy film with molten edges. Therefore, we formed p^+-p-n^+ structures by the alloying method. The p^+ contact was formed using an In + Cd alloy and the n^+ contact was formed from an In + Te alloy. It was found that the injection conditions were not affected when the concentrations of Cd and Te in the In + Cd and In + Te alloys were increased from 1 to 10%, as demonstrated by the current–voltage (I–V) characteristics plotted in Fig. 4. We determined the I–V characteristics for samples with the same geometry but in one case we used the In + 1% Te for the n^+ contact and the In + 1% Cd alloy for the p^+ contact; in the other case we used the In + 10% Te and In + 10% Cd alloys. We also studied samples with contacts formed using In + Te + Se and In + Cd + Zn alloys with different amounts of Te and Se, and Cd and Zn, but once again we found no significant changes in the I–V characteristics. Hence, we concluded that all the features of the I–V characteristics were associated with the p-type region.

The technology used in the preparation of the samples intended for studies of spontaneous radiation were slightly different from the technology used for the samples intended for an investigation of coherent emission.

The samples for the spontaneous emission study were prepared from unoriented special-grade p-type indium antimonide with initial impurity concentrations $p_0 = N_A - N_D = 2 \cdot 10^{13}$ cm^{-3} and $p_0 = N_A - N_D = 10^{16}$ cm^{-3}.

Our InSb platelets were ground to the required thickness (0.7-0.1 mm) and were then etched in a CP-4 polishing solution; a platelet was then broken up into pieces of required shape. The alloying of the contacts was carried out either simultaneously or in turn: first the n^+ contact and then the p^+ contact. The maximum areas of the samples prepared in this way were $\sim 10^{-1}$ cm^2 and the minimum areas were 10^{-3} cm^2. It was very difficult to prepare samples of larger cross section by the alloying method. These difficulties resulted from the low melting point of InSb, which was 560°C [67]. The melting points of the In + Te and In + Cd alloys were within 150°C. Fast heating and cooling were required to form abrupt junctions [68]. Optimal alloying conditions and the optimal thickness of the alloy layer were selected for each thickness of the sample so as to avoid fusion of the whole sample during the alloying operation. The

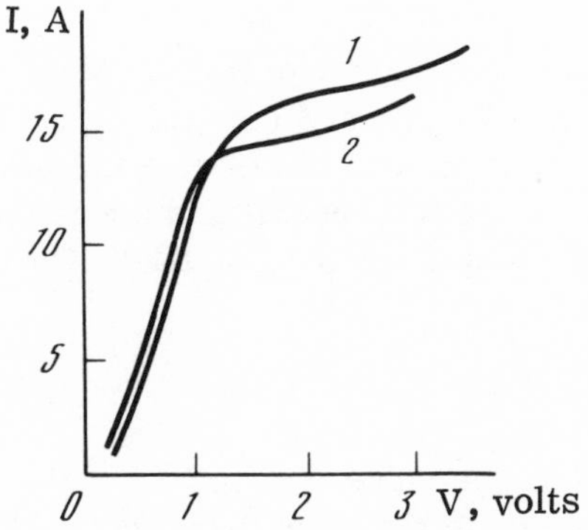

Fig. 4. Current–voltage (I–V) characteristics of samples prepared by alloying with 1% Te + In and 1% Cd + In (1) and with 10% Te + In and 10% Cd + In (2).

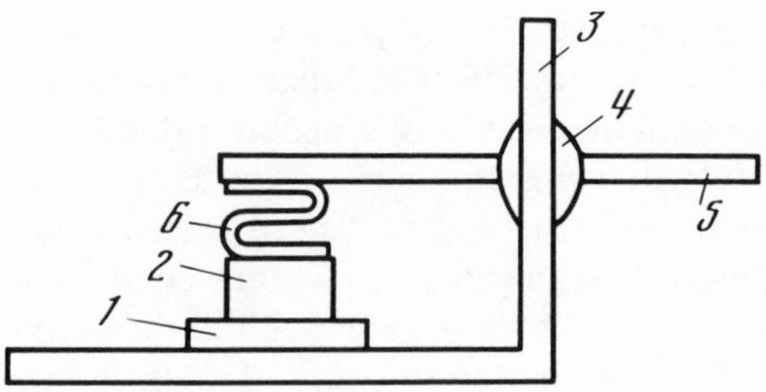

Fig. 5. Diode with holder: 1) copper spacer; 2) InSb sample; 3) copper bracket; 4) epoxy resin; 5) electrode; 6) copper strip.

alloying temperatures varied from 180°C for samples with the p-type region $L \leq 100\ \mu$ thick to 280°C for samples with $L \gtrsim 400$–$700\ \mu$. Alloying was followed by cleaving until the area of the sample became equal to the areas of the upper and lower contacts and then the samples were etched electrolytically until the required thickness was obtained; etching also reduced the surface recombination velocity.

The preparation of the samples intended for the observation of coherent radiation was different during the first stage (preparation of the p-type region before alloying) and it also differed by the absence of additional treatment after alloying.

The presence of reflecting faces (resonators) was essential for the observation of coherent radiation. Cleaving provided a convenient method for forming such reflecting faces. As in the case of other III-V compounds, cleaving of indium antimonide was easiest along the (110) planes [69], but it was also quite easy along (100) planes. Rectangular samples were obtained by cleaving when p-type InSb crystals were oriented with the aid of the Laue diffraction patterns (obtained with a URS-50 diffractometer); the orientation was accurate to within 1-2°. The crystal was then cut into plates 0.6-0.8 mm thick parallel to the (110) plane and then one of the sides of the plate was oriented exactly by additional polishing and a check with the aid of an x-ray diffraction goniometer to within at least 10'. Next, the opposite face was ground down to give a thickness of 0.15-0.5 mm. A plate obtained in this way was split into rectangular samples and the two ground surfaces were used for the alloying of contacts. This sample preparation method ensured a low surface recombination velocity and, therefore, further reduction in this velocity as a result of electrolytic etching was slight. The samples used in a study of coherent emission were prepared in the same way as the other samples but the contact areas were not always equal to the cross-sectional area of the sample.

Carrier heating by an electric field was minimized by reducing the drift component of the current in the sample. This was achieved by making the length of the p-type region equal to or shorter than the ambipolar diffusion length at high injection rates. We prepared samples with the p-type region 0.7-0.06 mm long. A study of the I–V characteristics indicated that the condition $L \leq L_a$ was obeyed in the $L < 450\ \mu$ range; here, L_a is the ambipolar diffusion length (see Chap. III, § 1 below).

Heat removal was improved by attaching a sample to a copper bracket. Alloying was then carried out in two stages. First, the n^+ contact was formed and then the p^+ contact was deposited on copper and the alloying was performed. This method of alloying the lower contact avoided the formation of a layer of pure indium between the crystal and copper holder (pure indium is a poor heat conductor). The construction of a sample together with its holder is shown in Fig. 5.

§ 2. Experimental Method

Large amounts of carriers were injected by passing a current of up to $j \approx 10^5$ A/cm^2 density through a sample. In view of the finite dimensions of the sample, it was necessary to

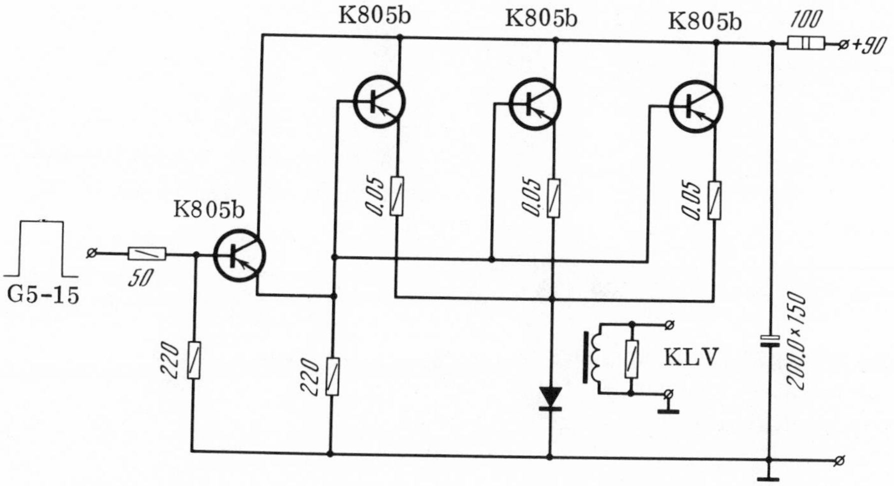

Fig. 6. Circuit of a rectangular pulse generator (amplifier). Here, G5-15 is a master pulse generator and KLV is a vacuum-tube voltmeter.

pass large currents (up to 30 A), which carried considerable power. This could have caused heating because of the Joule losses (I^2R) resulting from the finite resistance of the sample. The heating effect could be minimized by injecting the current in the form of pulses.

Pulse injection of high currents required a high-power pulse generator. The series resistance of our samples did not exceed 0.2-1 Ω at high injection rates. The available pulse generators had outputs of much higher resistance and then the main power was delivered to the resistance and not to the sample; i.e., the power dissipated in a sample was considerably less than 80 W. High currents were generated with a transistor pulse generator (amplifier) [70]. The low input resistance of A-406 transistors (~1 Ω) made it possible to achieve currents up to 35 A when the load resistance was 1 Ω. This generator had the circuit shown in Fig. 6.

The heating of the sample was reduced by ensuring that the pulses had sufficiently steep edges. Good rectangular shape of the pulses, i.e., the best frequency characteristic, was ensured by the common-emitter configuration of the transistor circuit (Fig. 6).

The transfer factor was determined as a function of the load resistance (Fig. 7). Even when the load resistance was only 0.2 Ω, the transfer factor was ≈70%, i.e., the system had a reasonable frequency characteristic.

The master pulse generator was of the G5-15 type. The current passing through the sample was measured with a pulse transformer, whose primary winding was the connection to the sample. The voltage in the secondary winding, calibrated relative to the current in the primary circuit, was measured with a VK-4 pulse voltmeter. The transformer had a ferrite core 20 mm in diameter and its magnetic permeability was μ = 600. The transformer measurement of the current was a satisfactory procedure in the case of relatively short pulses. When

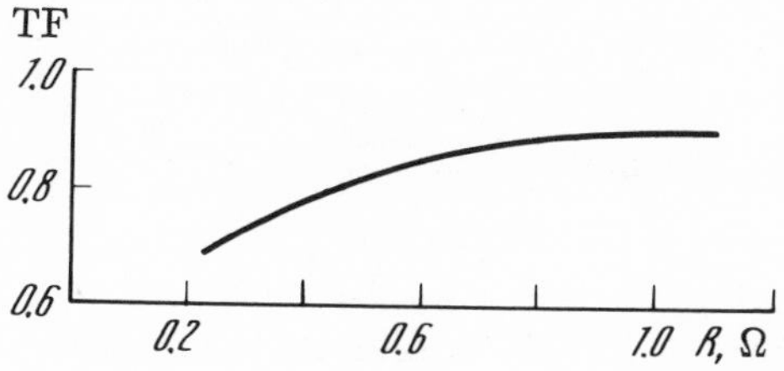

Fig. 7. Dependence of the transfer factor on the load resistance.

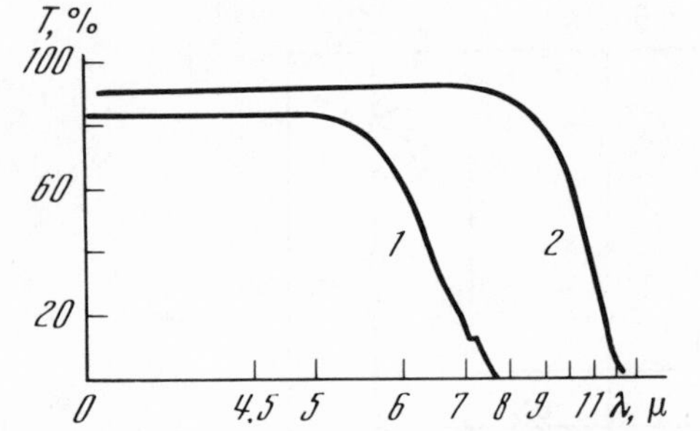

Fig. 8. Dependences of the transmission coefficient of sapphire (1) and fluorite (2) windows on the wavelength, determined at T = 300°K.

the pulses were longer or the currents were higher, there could be distortions in the measured current pulses. However, we found no deviation from Ohm's law when the current was determined as a function of the voltage up to I = 30 A using a variety of loads and pulse durations.

Certain difficulties were encountered in the determination of the emission spectra of InSb samples at 4.2 and 20°K. At these temperatures a sample with its holder was immersed directly in a helium or hydrogen bath and the emission wavelength was within the range from 5.9 to 5.06 μ. It was difficult to find optical windows which would have a good transmission coefficient, high optical strength, and low thermal expansion coefficient in this range of wavelengths. Radiation was extracted from the helium bath via sapphire windows whose transmission spectrum was determined (curve 1 in Fig. 8). We found that in the wavelength range $\lambda < 5\ \mu$ the transmission coefficient was close to 80% whereas in the $\lambda \gtrsim 5.5\ \mu$ range this coefficient fell to 70%. The outer windows (kept at T = 300°C) were made of fluorite (CaF_2), whose transmission coefficient in the same range of wavelengths was constant and equal to 90% (the transmission spectrum of CaF_2 is represented by curve 2 in Fig. 8).

We used a helium cryostat shown in Fig. 9. Electrical power was supplied to a sample by a coaxial waveguide in the form of two stainless steel tubes. These thin-walled tubes were

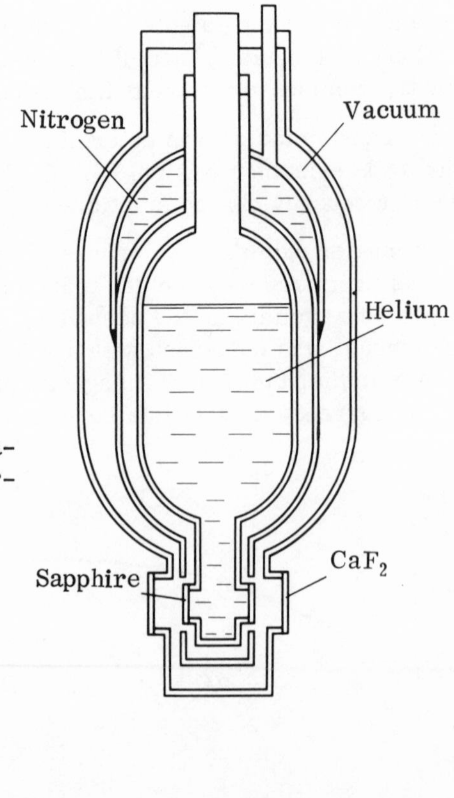

Fig. 9. Optical cryostat for the determination of the emission spectra at low temperatures.

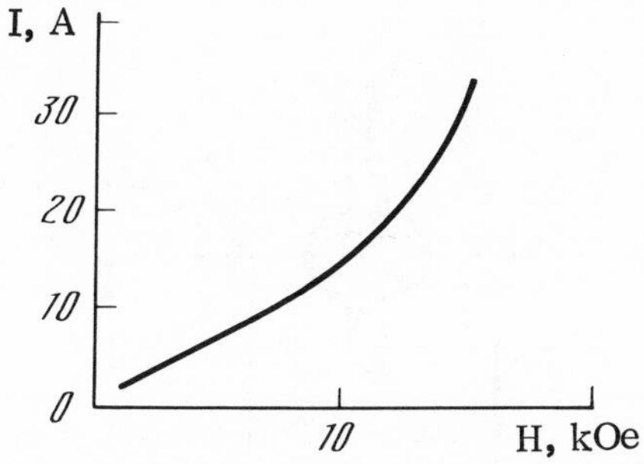

Fig. 10. Calibration curve giving the dependence of the magnetic field intensity on the current through the electromagnet winding.

used as the conductor because of the need to reduce heat transfer between liquid helium and the pulse generator which was at room temperature. This cryostat had a special stem because some measurements of the electrical conductivity and emission spectra were carried out in a magnetic field. A field H up to 15 kOe was produced by an electromagnet. We determined the dependence of the magnetic field on the current flowing through the electromagnet winding (Fig. 10). This magnetic field was produced in a gap ~42 mm wide by polepieces ~60 mm in diameter. When the dimensions of the samples were ~0.7-0.1 mm, the magnetic field was homogeneous; its value was deduced from the calibration curve in Fig. 10.

The radiation emitted from a sample in this cryostat was focused by a CaF_2 lens onto the slit of an IKM-1 monochromator with a replica grating. The use of a replica ensured a line resolution of $\sim 5 \cdot 10^{-5}$ eV (in the stimulated emission regime) when the monochromator slits were 50-30 μ wide. It was undesirable to use narrower slits because the emission wavelength was 5 μ and a diffraction could then be observed. This lens produced a magnified (by a factor of 1.5) image of the sample on the monochromator slit. Thus, in the stimulated emission regime, when the slits were not more than 50-30 μ wide, we observed the radiation emitted from a part of a crystal 35-20 μ. The radiation spectrum passed through the monochromator and was then focused by an NaCl lens on a radiation detector. This detector was an InSb photoresistor, a Ge : Au photoresistor, or an InSb photodiode cooled to 77°K. The InSb photoresistor had the highest sensitivity in the 5 μ wavelength range, but the InSb photodiode had better time characteristics and the constancy of the spectral sensitivity of the Ge : Au photoresistor throughout the investigated range made it possible to detect any distortions in the recombination radia-

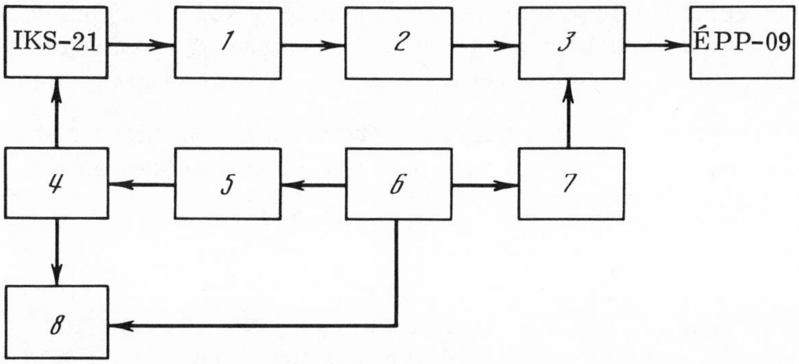

Fig. 11. Block diagram of the system used in the determination of the recombination radiation spectra: 1) InSb photoresistor; 2) amplifier; 3) synchronous detector; 4) InSb crystal; 5) pulse generator; 6, 7) G5-15 generator; 8) double-beam oscillograph.

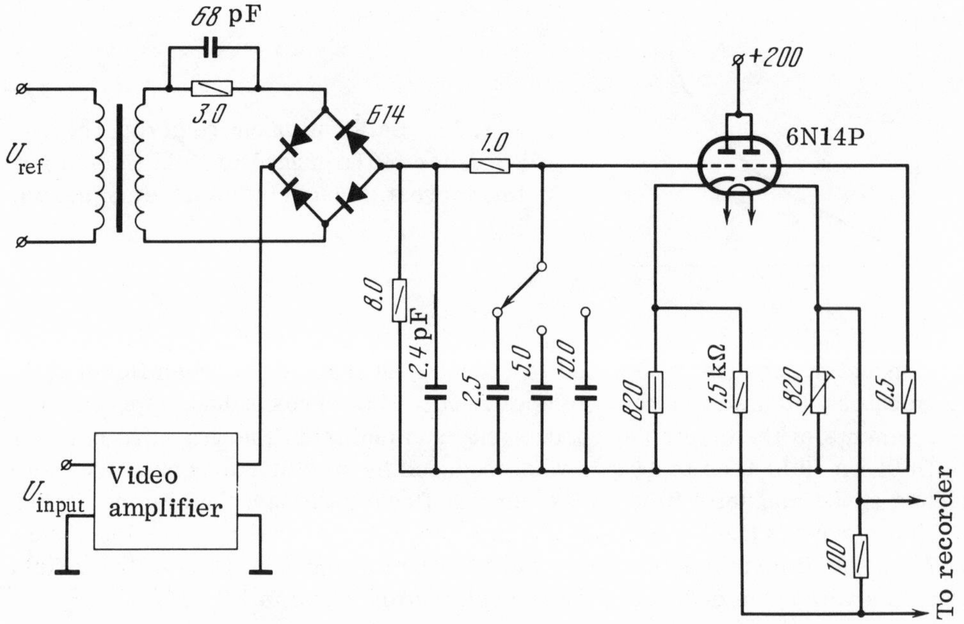

Fig. 12. Synchronous detector used in recording recombination radiation spectra.

tion spectrum due to a reduction in the transmission coefficient of sapphire and the sharp absorption edge of the InSb detectors.

It was found that the recombination radiation spectrum was not greatly distorted by the spectral characteristic of the InSb detectors even when the measurements were carried out in the $\lambda \approx 5.9\ \mu$ range.

A signal from the detector was applied to a USh-2 wide-band amplifier and then to a synchronous detector; it was recorded with an ÉPP-09 automatic potentiometer. We recorded the recombination radiation spectra using the system shown schematically in Fig. 11. Synchronous detection was used to improve the detector resolution when the photoresistor signal was low. The principle of the operation of the system was that a signal was detected only during a time interval equal to the duration of a reference pulse. Synchronization between the current and reference pulses was achieved by synchronous triggering of the G5-15 pulse generators. The accumulation time constant varied from 5 to 50 sec. The use of a synchronous detector circuit improved considerably the signal/noise ratio. This circuit was of the type shown in Fig. 12.

The detector had a good linearity beginning from the lowest signals and was sensitive to the sign of the applied pulses. The amplitude of the reference pulse was [70]

$$\frac{E}{R_L}\left[(R_r + R_g)\frac{T}{\tau} + R_L\right] = U_L, \qquad (31)$$

where E is the static voltage in the circuit; T is the pulse repetition period; τ is the duration of the reference pulse; R_r and R_g are the resistance of the sample under a forward bias and the internal resistance of the pulse generator; R_L is the load resistance; U_L is the amplitude of the reference signal across this load resistance.

Using a graph given in [71], we were able to determined the transfer factor of the diode detector. Under our conditions this factor was 0.6-0.7. The amplitude of the reference signal was an order of magnitude greater than the amplitude of the signal pulses.

Measurements of the pulse I–N characteristics were carried out using an S1-7 oscillograph. The same oscillograph was used in estimating the time characteristics of the carriers from the shape of the optical output and current input pulses.

CHAPTER III

RECOMBINATION RADIATION EMITTED FROM INDIUM ANTIMONIDE AT LOW TEMPERATURES

Among the currently known III-V noncentrosymmetric compounds, only InSb can be regarded as a direct-gap semiconductor right down to T = 4.2°K because this compound is characterized by the smallest shift of the heavy-hole valence band in the ⟨111⟩ direction relative to **k** = 0.

The characteristic features of the energy band structure of InSb, namely the large difference between the density-of-states effective masses of electrons and holes ($m_1/m_2 \approx 40$) and the complex structure of the valence band (the presence of light and heavy holes), permit two radiative transitions (electron–light hole and electron–heavy hole) and govern the nature of the recombination radiation spectra.

The energy band structure of InSb at two rates of injection is shown schematically in Fig. 13. The shaded regions correspond to electron occupancy. The numbers 1 and 2 are used to denote the electron–light hole and electron–heavy hole transitions. It is clear from Fig. 13 that, because of the large difference between the effective masses, the emission spectrum due to the electron–heavy hole (2) radiative transitions at T = 0 is indeed governed by the electron energy distribution in the conduction band. The nature of the emission spectrum due to the electron–light hole radiative transitions depends not only on the electron but also on the light-hole energy distributions, and if the quasi-Fermi level of electrons F_e becomes greater than kT ($n \approx 10^{14}$ cm^{-3}), this spectrum is governed only by the light-hole energy distribution.

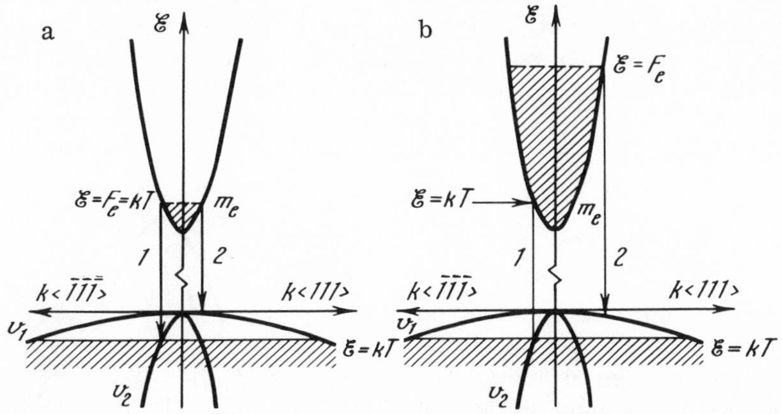

Fig. 13. Energy-band scheme of InSb plotted for two rates of injection without allowance for the valence band nonparabolicity or for the shift of the heavy-hole band maximum in the ⟨111⟩ direction: a) low rate of injection; b) high rate of injection.

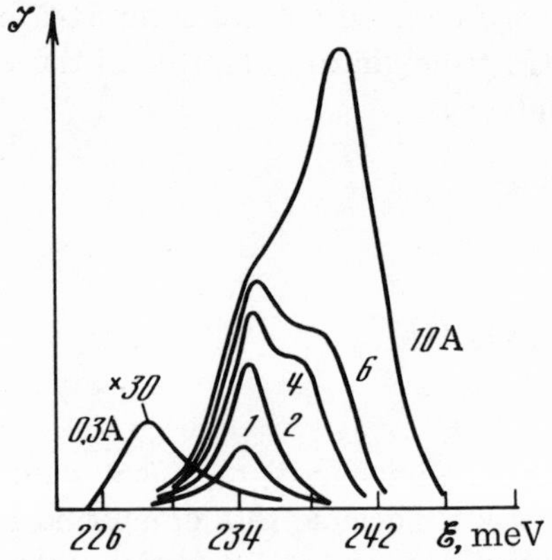

Fig. 14. Recombination radiation spectra obtained for pure indium antimonide crystals at T = 4.2°K using different injection currents given alongside each curve.

§ 1. Characteristics of Recombination Radiation Emitted by Pure InSb Crystals at 4.2°K

It follows from the absorption data obtained for relatively pure and doped n- and p-type indium antimonide crystals that the absorption coefficient for the transitions from the light-hole to the conduction band depends on energy as follows:

$$\alpha(h\nu) \propto \sqrt{h\nu - \mathcal{E}_g}. \tag{32}$$

In the case of transitions from the heavy-hole to the conduction band the absorption coefficient rises with energy much more slowly than predicted by Eq. (32).

A comparison of the experimental results obtained by Gobeli and Fan [12] for pure InSb crystals with the dependences (32) and

$$\alpha(h\nu) \propto (h\nu - \mathcal{E}_g)^{3/2} \tag{33}$$

was made by Johnson [11] and it demonstrated that the law (32) described better the experimental results. Gobeli and Fan suggested that the slow rise of the absorption coefficient in the case of transitions from the heavy-hole to the conduction band was due to a strong anisotropy of the heavy-hole band and they estimated the degree of this anisotropy. Since the luminescence was the opposite to the absorption, one could expect that in a certain range of the injected-carrier densities the intensity of the electron–heavy hole radiative transitions would not be

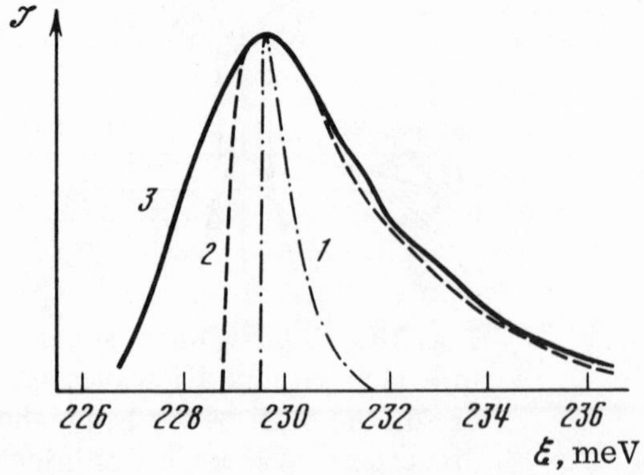

Fig. 15. Spectrum representing electron–acceptor level radiative transitions in the case of low injection rates at T = 4.2°K: 1) theoretical curve calculated in [72] for T = 4.2°K; 2) theoretical curve calculated in [72] for T = 20°K; 3) experimental spectrum.

proportional to $\sqrt{\mathscr{E}}$ because of the slower than $\rho \propto \sqrt{\mathscr{E}}$ ($\mathscr{E} = h\nu - \mathscr{E}_g$) variation of the density of states (ρ) in the case of the direct electron–heavy hole optical transitions.

Figure 14 shows the recombination radiation spectra obtained for samples with hole densities $p_0 = N_A - N_D = 2 \cdot 10^{13}$ cm^{-3} as a result of injection at different rates.

a) Low Injection Rates. It is clear from Fig. 14 that at low injection rates (curve corresponding to I = 0.3 A), when the carrier density is $n \ll p_0$, the recombination radiation spectra are dominated by the transitions of electrons to the zinc acceptor level. The energy corresponding to the maximum of this transition $\mathscr{E}_{max} \approx 229$ meV should be equal to the energy gap $\mathscr{E} = \mathscr{E}_g - \mathscr{E}_I + kT$, as demonstrated by Eagles [72], who calculated the form of the recombination spectrum for electron–acceptor level transitions allowing for the Boltzmann distribution of electrons in the conduction band. Figure 15 shows the recombination radiation spectrum for electron–acceptor level transitions determined at low injection rates as well as the theoretical curves plotted using the formula $\text{Int}\, y = \sqrt{y} \exp(-y)$, where

$$y = \frac{h\nu - \mathscr{E}_g - \mathscr{E}_I}{kT}, \tag{34}$$

at two temperatures of 4.2 and 20°K. It is clear from this figure that the long-wavelength part of the spectrum is described poorly by the Eagles formula but the short-wavelength range agrees with the curve calculated for T = 20°K.

In his calculations, Eagles assumed an isolated impurity level and did not allow for excited states of the impurity center. Allowance for the finite width of the impurity level and for the excited states should broaden the theoretical spectra at both temperatures.

Thus, the nature of the recombination radiation spectrum representing the electron–acceptor level transitions shows that at low injection rates the electron temperature is higher than 4.2°K but lower than 20°K; however, a more accurate estimate of the temperature of the electron gas from the recombination radiation spectra cannot be obtained without allowance for the width of the impurity level.

b) High Injection Rates. When the injected carrier density is $n = p \gg p_0 = N_A - N_D$, the recombination radiation spectra of indium antimonide have two maxima (Fig. 14) at photon energies close to the forbidden band width $\mathscr{E}_g = 235.7$ meV [73]. The energy position and the half-width of the long-wavelength maximum are practically unaffected by the carrier density right up to high injection rates. The energy and half-width of the short-wavelength maximum increase with the injected-carrier density, as shown in Fig. 14. The dependences of the energy positions and half-widths of these maxima on the rate of injection fit well the hypothesis of radiative transitions: electron–light hole and electron–heavy hole, as indicated in Fig. 13. It is clear from Fig. 13 that the energy of a maximum due to the electron–light hole transitions is governed by the energy distribution of the light holes. As the rate of injection increases right up to a value such that the injection-carrier density corresponds to degeneracy in the valence band, the maximum of the light-hole density shifts to an energy ~kT and the maximum of the electron density becomes located close to the Fermi energy F_e. Therefore, the energy of the maximum due to the electron–light hole transitions is close to $\mathscr{E} = h\nu - \mathscr{E}_g \approx 2kT$. In the case of the electron–heavy hole transitions the energy at the maximum is close to the Fermi level F_e of electrons and the spectrum is governed by the electron energy distribution in the conduction band if we ignore the anisotropy of the heavy-hole valence band.

Using the results from Refs. 11 and 12, we can describe the dependences of the recombination radiation intensity on the energy for the two transitions in the form

$$\text{Int}(\mathscr{E}) = A_1 \sqrt{\mathscr{E}}\, f_e(\mathscr{E}) f_1(\mathscr{E}) \text{ for transitions 1,} \tag{35}$$

$$\mathrm{Int}(\mathscr{E}) = A_2\mathscr{E}^{3/2}f_e(\mathscr{E})f_2(\mathscr{E}) \text{ for transitions 2,} \tag{36}$$

where $\mathscr{E} = h\nu - \mathscr{E}_g$.

We have to allow for the fact that in the case of the electron–light hole radiative transitions the transition energy measured from $h\nu - \mathscr{E}_g$ increases with the wave vector twice as fast as for the electron–heavy hole transitions, because the effective masses of electrons and light holes are practically identical, i.e., $m^* = m_e = m_1 = 0.014m_0$, but the effective mass of heavy holes is much greater.

The electron and light-hole distribution functions $f_e(\mathscr{E})$ and $f_1(\mathscr{E})$ in Eq. (35) are the Fermi–Dirac distribution functions $f(\mathscr{E}) = [1 + \exp(\mathscr{E}/2kT - \eta_i)]^{-1}$ $(\eta_i = F_i/kT)$. The factor of 2 appears because for the same effective masses of electrons and light holes, the photon energy $\mathscr{E} = h\nu - \mathscr{E}_g$ is twice as high as the carrier energies in the bands. The energy of the Fermi level of carriers in the conduction band is given by the usual formula

$$\eta_e = \frac{F_e}{kT} = \frac{\hbar^2}{2m_e}\left(\frac{3n}{8\pi}\right)^{2/3}/kT$$

in the $F_e \gg kT$ case and the quasi-Fermi level of holes has to be taken at a finite temperature. The electron and heavy-hole distribution functions $f_e(\mathscr{E})$ and $f_2(\mathscr{E})$ in Eq. (36) are also the Fermi–Dirac functions. Since in the case of this direct transition 2, we have $\mathscr{E}_{\max} = h\nu - \mathscr{E}_g \simeq F_e$, the electron distribution function is

$$f_e(\mathscr{E}) = \left[1 + \exp\left(\frac{\mathscr{E}}{kT} - \eta_e\right)\right]^{-1},$$

and the hole distribution function is

$$f_2 = \left[1 + \exp\left(\frac{m_e}{m_p}\frac{\mathscr{E}}{kT} - \eta_p\right)\right]^{-1}.$$

In calculating the form of the recombination radiation spectra in the case of the electron–heavy hole transitions it is necessary to allow for the anisotropy of the heavy-hole band (Fig. 16). This anisotropy not only results in a much lower rise of the absorption coefficient in the range $\mathscr{E} < F_e$, but also in a slower fall in the range $\mathscr{E} > F_e$. The electron distribution function is defined by

$$f_e(\mathscr{E}) = \left[1 + \exp\frac{\eta - \eta_e}{2.6}\right]^{-1} \simeq \left[1 + \exp\frac{\eta - \eta_e}{3}\right]^{-1};$$
$$\eta = \frac{\mathscr{E}}{kT}.$$

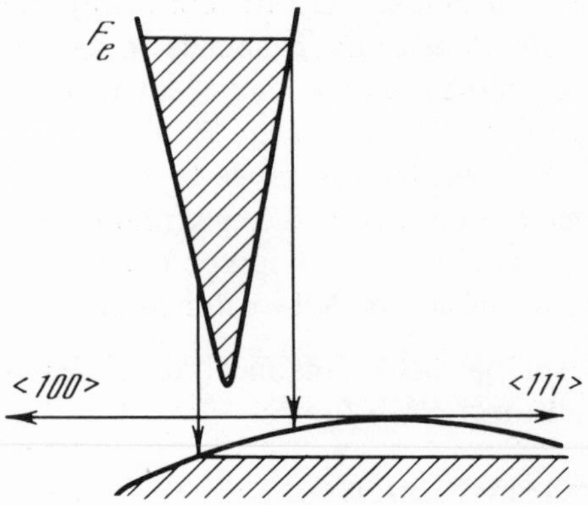

Fig. 16. Scheme showing electron–heavy hole radiative transitions plotted allowing for the nonparabolity of the heavy-hole valence band and for the shift of its maximum in the ⟨111⟩ direction relative to **k** = 0.

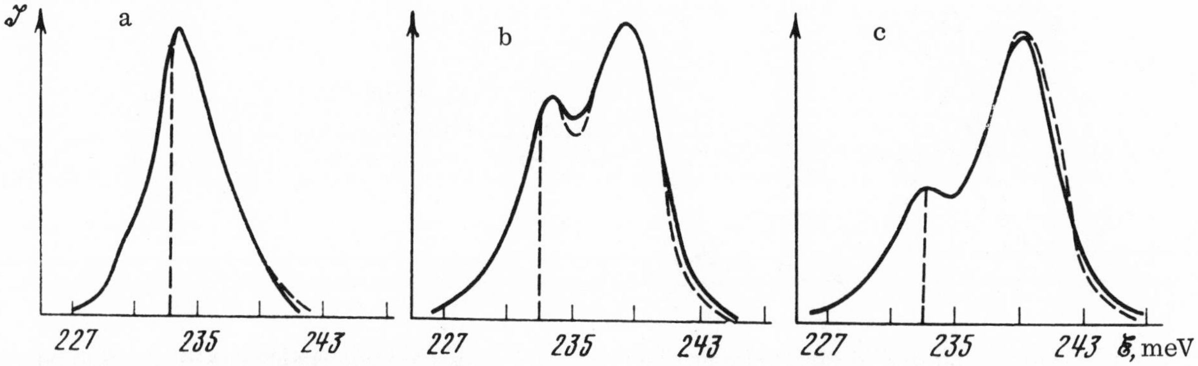

Fig. 17. Recombination radiation spectra of pure indium antimonide ($p_0 = 2 \cdot 10^{13}$ cm^{-3}) obtained at T = 4.2°K using different injection rates: a) $n = p = 5 \cdot 10^{14}$ cm^{-3}; b) 10^{15} cm^{-3}; c) $8 \cdot 10^{15}$ cm^{-3}; the dashed curves are the theoretical spectra calculated using Eqs. (37) and (38).

The final forms of the dependences of the recombination radiation intensity on the energy are

$$\mathrm{Int}(\eta) = \sqrt{\eta} A_1 \left[1 + \exp\frac{\eta - \eta_e}{2}\right]^{-1} \times \left[1 + \exp\frac{\eta - \eta_p}{2}\right]^{-1}, \quad (37)$$

$$\mathrm{Int}(\eta) = \eta^{3/2} A_2 \left[1 + \exp\frac{\eta - \eta_e}{3}\right]^{-1} \times \left[1 + \exp\left(\frac{m_e}{m_p}\eta - \eta_p\right)\right]^{-1}. \quad (38)$$

The results of calculations of the recombination radiation spectra for the currents I = 4, 6, and 10 A are plotted in Fig. 17. In these calculations it is assumed that F_e = 12kT, 17kT, and 24kT for the currents I = 4, 6, and 10 A, respectively, which corresponds to the condition $I \propto n$. A good agreement with the experimental spectra is obtained if the ratio of the constants A_1 and A_2 in the summation of the heavy- and light-hole spectra is assumed to be $A_1 : A_2 = 30$ and the energy at the bottom of the valence band is taken to be 233 meV. The rates of injection between F_e = 12kT and F_e = 24kT (T = 4.2°K) correspond to carrier densities from $2 \cdot 10^{15}$ to $8 \cdot 10^{15}$ cm^{-3}. When the electron density is $n \approx (2\text{-}4) \cdot 10^{15}$ cm^{-3}, the change in the forbidden band width because of the exchange interaction energy of electrons is $\mathscr{E}_{\text{exch}} = 2.12/r_s$ Ry [74] = 2.4-2.7 meV, i.e., the forbidden band width should be 233 meV, as allowed for in our calculations. It is clear from Fig. 17 that the calculated and theoretical curves agree well for the currents 4 and 6 A, but the agreement is poorer for 10 A. Since for I = 10 A, we have $n \approx 8 \cdot 10^{15}$ cm^{-3} and $\mathscr{E}_{\text{exch}}$ = 3.3 meV, and the average electric field in the sample is higher, it follows that dropping of the bottom of the conduction band because of the exchange interaction energy and also because of the influence of the Franz–Keldysh effect has to be allowed for in the calculation of the spectrum for I = 10 A. The value of 30 for the quantity $A_1/A_2 = (|M_1|^2/|M_2|^2)[\rho_1(k)/\rho_2(k)]$ is clearly due to the higher density of states ρ_1 for the transitions of type 1 compared with the density of states ρ_2 for the transitions of type 2.

§ 2. Redistribution of Carriers in a Magnetic Field between the Light- and Heavy-Hole Valence Subbands

An investigation of the Shubnikov–de Haas oscillations in heavily doped indium antimonide samples, carried out by Bir, Parfen'ev, and Tamarin [37], demonstrated that the application of a magnetic field altered the probabilities of the interband scattering between the light- and heavy-hole subbands if the magnetic field was such that quantization at the Landau

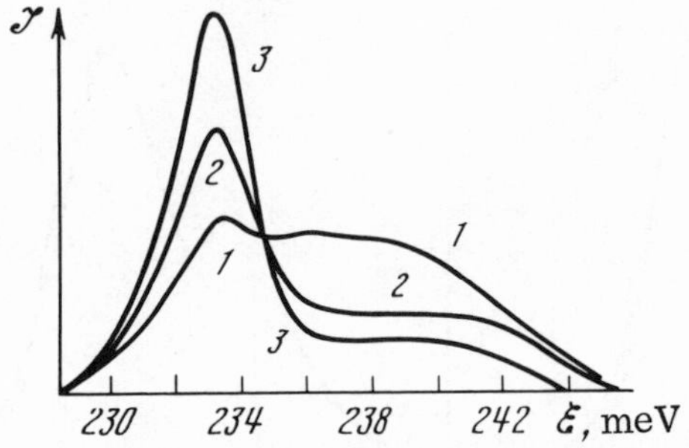

Fig. 18. Changes in the recombination radiation spectrum of a sample with $p_0 = 2 \cdot 10^{13}$ cm^{-3} carrying a current I = 10 A at 4.2°K: 1) H = 150 Oe; 2) H = 200 Oe; 3) H = 350 Oe.

level was observed in the light-hole subband and the heavy-hole subband was still unaffected by the magnetic field. The change in the probability of the interband transitions in a magnetic field was so great (from $w_{12} > w_{21}$ for H = 0 to $w_{12} < w_{21}$ for $H \neq 0$) that it was natural to expect a change in the recombination radiation intensity in the case of the electron–light hole and electron–heavy hole transitions in a magnetic field.

Figure 18 shows the recombination radiation spectra obtained for a current of I = 12 A in magnetic fields H = 150, 200, and 350 Oe. Clearly, an increase in the magnetic field resulted in a redistribution of the recombination radiation intensity between two maxima in favor of the maximum corresponding to the electron–light hole transitions [75]. This redistribution of the radiation intensity was observed at practically any of the investigated injection rates.

Figure 19 shows the recombination radiation spectra obtained in magnetic fields H = 0 and 500 Oe using currents of 2, 10, and 21 A.

The application of a magnetic field resulted in a separation of the long-wavelength maximum corresponding to the electron–light hole transitions, and the intensity of this maximum

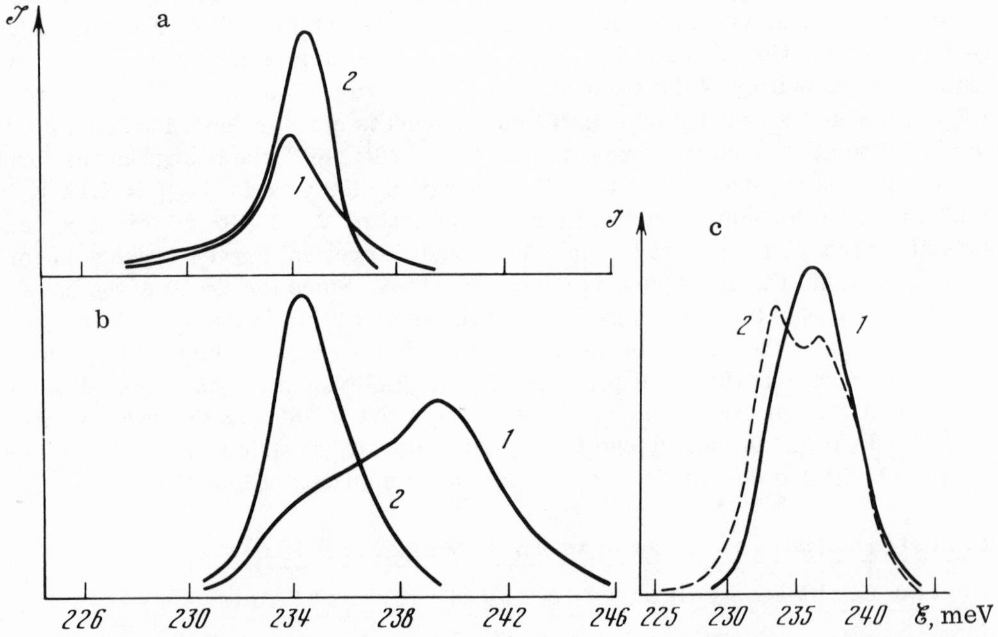

Fig. 19. Recombination radiation spectra of a sample with $p_0 = 2 \cdot 10^{13}$ cm^{-3} obtained using different currents and different magnetic fields: a) I = 2 A; b) I = 10 A; c) I = 21 A; 1) H = 0; 2) H = 500 Oe; T = 4.2°K.

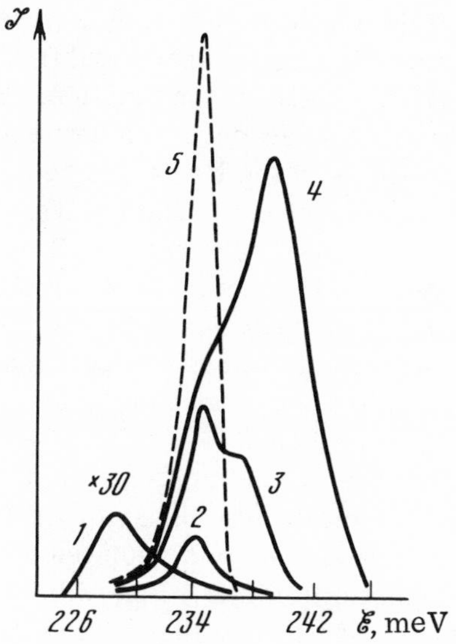

Fig. 20. Recombination radiation spectra of pure InSb obtained at different injection rates in the absence of a magnetic field (1-4) and in a magnetic field H = 500 Oe (5) at T = 4.2°K: 1) I = 0.3 A; 2) I = 2 A; 3) I = 4 A; 4) I = 10 A; 5) I = 4 A.

increased linearly with the rate of injection. This was evidence of a linear recombination law, which should apply to the electron–light hole transitions provided the Fermi level of electrons was $F_e \gg kT$.

In a magnetic field H = 500 Oe for a current I = 10 A there was not only an increase in the intensity of the maximum corresponding to the electron–light hole transitions but also an increase in the total intensity compared with that obtained in H = 0 due to the electron–light hole transitions, as demonstrated in Figs. 19 and 20. At high injection rates (I = 21 A) the recombination radiation intensity in a magnetic field H = 500 Oe was only slightly higher than in H = 0, although the intensity of the electron–light hole line increased strongly. This behavior of the intensity in the magnetic fields was due to the finite density of states at the n = 0 Landau level in the light-hole band in a magnetic field H = 500 Oe. An exact calculation of the density of states at the n = 0 Landau level in the light-hole band is difficult or even impossible because of the complex structure of the valence band [14, 15]. The probability of carrier transfer be-

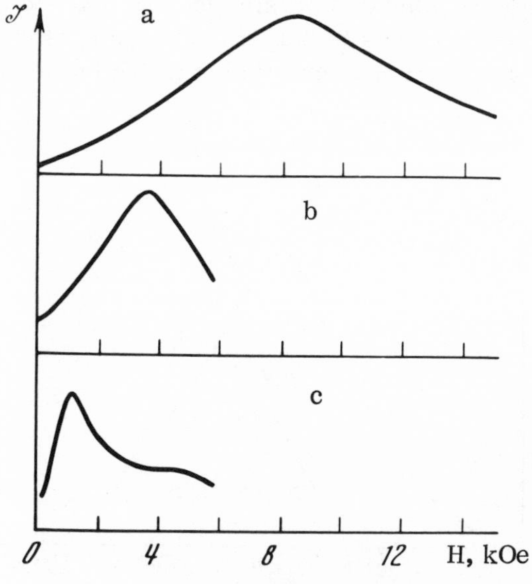

Fig. 21. Magnetic-field-induced changes in the integrated intensity of the recombination radiation emitted from samples with different thicknesses (lengths) of the p-type region; a) 450 μ; b) 250 μ; c) 60 μ.

tween the light- and heavy-hole subbands in a magnetic field and in its absence is governed by the matrix element of the transition and by the carrier densities in the initial and final states, i.e., it depends on the density of states and the carrier energy distribution function. In a given magnetic field the distribution of holes between the subbands corresponds to a new equilibrium condition governed by a different ratio of the densities of states in the light- and heavy-hole subbands. Figure 21 shows how the intensity of the electron–light hole radiative transitions changes with the length of the p-type region. It is clear from this figure that in all cases the recombination radiation intensity has a maximum but these maxima occur in different magnetic fields, which depend on the length of the p-type region. In the longest sample the intensity maximum is observed in a magnetic field H = 8 kOe, in the sample whose p-type region is 250 μ long it occurs in a field of ~4 kOe, and in the sample with the corresponding length of 60 μ it occurs in ~1 kOe. We shall not consider here the factors responsible for the dependence of H_{max} on the length of the crystal but we shall mention that the recombination radiation intensity passes through a maximum at a photon energy of 235 meV. The behavior of the intensity of the electron–light hole line in magnetic fields is governed by a redistribution of holes between the light- and heavy-hole subbands. In magnetic fields $H < H_{max}$ the density of states rises at the ligh-hole n = 0 Landau level and at the same time the level shifts into the valence band. When the Landau level energy exceeds kT in the valence band ($H > H_{max}$), carriers return back to the heavy-hole band.

§ 3. Energy of Electron – Light Hole Radiative Transitions in Strong Electric and Magnetic Fields

The greatest influence of an external electric field on carriers in p^+-p-n^+ structures is observed in those samples which have p-type regions of length comparable with or greater than the ambipolar diffusion length (see Chap. IV, § 1 below). Figure 22 shows the recombination radiation spectra for samples with p-type regions L = 450 μ and L = 250 μ long, recorded at low injection rates. An increase in the length of the p-type region clearly shifts the recombination radiation maximum toward longer wavelengths, increases the width of the spectrum, and reduces the relative intensity of the recombination radiation emitted from the longer sample.

This shift of the recombination radiation spectra toward longer wavelengths may be due to the heating of a crystal and a reduction in the forbidden band width because of the Franz–Keldysh effect [76, 77]. However, at low injection rates ($I \leq 1$ A, V = 0.2 volts), the average power supplied to a crystal does not exceed $2 \cdot 10^5$ W. If the surface of the sample is 10^{-2} cm^2, the power that is transferred to liquid helium is $2 \cdot 10^{-3}$ W/cm^2. This is much less than the threshold of 0.5 W/cm^2 [78] at which the temperature of such a crystal begins to rise in liquid

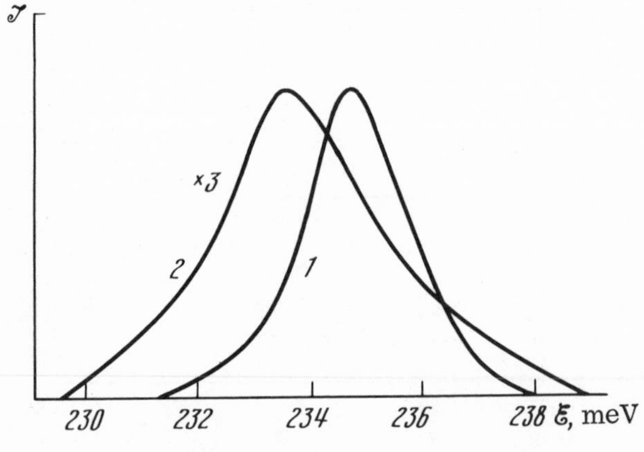

Fig. 22. Recombination radiation spectra recorded at low injection levels for two samples of pure indium antimonide with different thicknesses (lengths) of the p-type region at T = 4.2°K: 1) 0.25 mm; 2) 0.45 mm.

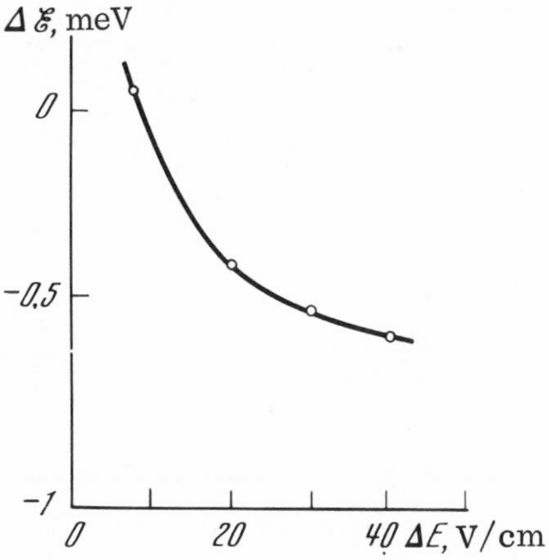

Fig. 23. Dependence of the energy at the maximum of the electron–light hole spectrum on the electric field applied to a sample 300 μ long.

helium. Therefore, the change in the energy of the recombination radiation maximum and in the width of the spectrum, exhibited by long p^+-p-n^+ structures, can only be due to the influence of the electric field on carriers.

A strong electric field is known to alter the absorption edge [79] because of the indeterminacy of the carrier energy in electric fields. This effect was predicted by Keldysh [76] and Franz [77] and it can be described as follows: the application of an electric field produces, at energies below the fundamental absorption edge, a density-of-states tail which decreases exponentially in the direction of longer wavelengths. An estimate of the forbidden band width in an electric field (a fall of the exponential function by a factor of e) obtained by Keldysh [76] shows that this change is given by

$$\Delta\omega = \frac{1}{2}(e\hbar^2E^2/m_r)^{1/3}. \tag{39}$$

Here, E is the electric field intensity and the reduced mass is given by $m_r = m_e m_p/(m_e + m_p)$, where the effective hole mass m_p is the light-hole mass m_1, as confirmed by the electroabsorption experiments carried out on semiconductors with a complex structure of the valence band [79]. Since the effective masses of electrons and light holes in indium antimonide are small, this effect should be considerable even in electric fields of a few hundreds of volts per centimeter, because $\Delta\omega = 2.5 \cdot 10^{-5}E^{2/3}$ eV for InSb. Thus, a change in the forbidden band width by 1 meV requires a field of the order of 300 V/cm. The same field gives rise to an indeterminacy in the thermal broadening of the light-hole distribution and this broadens the recombination radiation spectra.

Figure 23 gives the dependence of the change in the energy of the recombination radiation maximum, due to the electron–light hole transitions, on the applied electric field. The abscissa gives the electric field E relative to E = 100 V/cm, and the ordinate the change in the energy of this maximum relative to 234 meV. The measurements were carried out in a magnetic field H = 1 kOe, because in the absence of a magnetic field the electron–light hole radiative transition was not observed in electric fields $E \gtrsim 100$ V/cm. The electric field was assumed to be uniformly distributed across the thickness of the sample, which was true only of very long crystals and high fields. A considerable change in the photon energy due to a small change in the electric field ($E \lesssim 20$ V/cm) could be due to the inhomogeneity of the field distribution.

In a magnetic field H the carrier energy can be expressed in the form

$$\mathscr{E}(H) = \hbar^2\mathbf{k}^2/2m + \left(n + \frac{1}{2}\right)\hbar\omega_c \pm \frac{1}{2}g\beta H. \tag{40}$$

Here, $\hbar^2k^2/2m$ is the kinetic energy of a carrier in the magnetic field direction; $(n + {}^1/_2)\hbar\omega_c$ is the cyclotron interaction energy; $\pm{}^1/_2 g\beta H$ is the term associated with the spin splitting of the Landau level in a magnetic field. If $\hbar\omega_c > kT$, where $\hbar\omega_c = \hbar eH/mc$, there are singularities in the densities of states in the energy bands and these are due to the appearance of the Landau levels. The density of states in a magnetic field is

$$N(\mathscr{E}, H) = 2\pi\left(\frac{2m}{\hbar^2}\right)^{3/2}\sum_{n=0}^{\infty}\frac{\hbar\omega_c}{\left[\mathscr{E}-\left(n+\frac{1}{2}\right)\hbar\omega_c\right]^{1/2}}, \tag{41}$$

where n are integers from 0 to ∞.

The condition $\hbar\omega_c > kT$, in which the band splits into Landau levels, depends on the effective mass and it is satisfied in different magnetic fields by different carriers. In the case of electrons and light holes this condition is satisfied by magnetic fields $H \gtrsim 500$ Oe, whereas in the case of heavy holes it is satisfied in a field $H \gtrsim 15$ kOe. Therefore, in magnetic fields $H < 15$ kOe, we may assume that only the light-hole and electron bands are quantized. The Lande factor for electrons can be deduced from the formula [80]

$$g = 2\left[1-\left(\frac{m_0}{m_e}-1\right)\left(\frac{\Delta}{3\mathscr{E}_g+2\Delta}\right)\right], \tag{42}$$

where m_0 is the mass of a free electron; m_e is the effective electron mass; Δ is the spin–orbit interaction energy, which is 0.9 eV for InSb; $\mathscr{E}_g$ is the forbidden band width, equal to 236 meV. The results of the calculations and numerous experiments demonstrate that the value of the g factor of InSb is −50. Using $m = 0.014m_0$ and $g = -50$, we can calculate the change in the energy of the Landau level characterized by n = 0 and $m = \pm{}^1/_2$ in a magnetic field. In the investigated range of energies ($\Delta\mathscr{E} \approx 5$ meV from the bottom of the conduction band) the effective electron mass is essentially constant. The results of such a calculation of the n = 0 Landau level energy in a magnetic field and of the energy of the exciton level associated with this Landau level are plotted in Fig. 24. It follows from this figure that the electron energy at the Landau level varies with increasing H at the rate $\Delta\mathscr{E}/\Delta H = 0.26$ meV/kOe for $m_s = +{}^1/_2$ and at the rate $\Delta\mathscr{E}/\Delta H = 0.53$ meV/kOe for $m_s = -{}^1/_2$. The change in the exciton level energy in a magnetic field can be calculated following the method of Yafet, Keyes, and Adams [81]. We can see that in a field 2 kOe < H < 10 kOe, the change can be described by straight lines

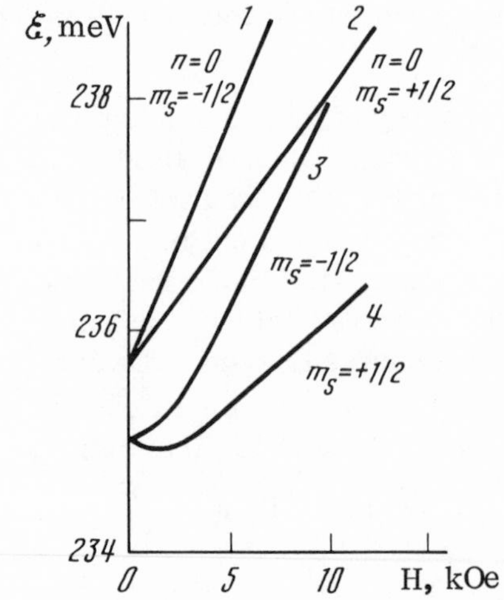

Fig. 24. Magnetic-field-induced changes in the electron energy at the Landau level with n = 0 and $m_s = \pm{}^1/_2$ (curves 1 and 2) in the energy of the associated exciton level (curves 3 and 4).

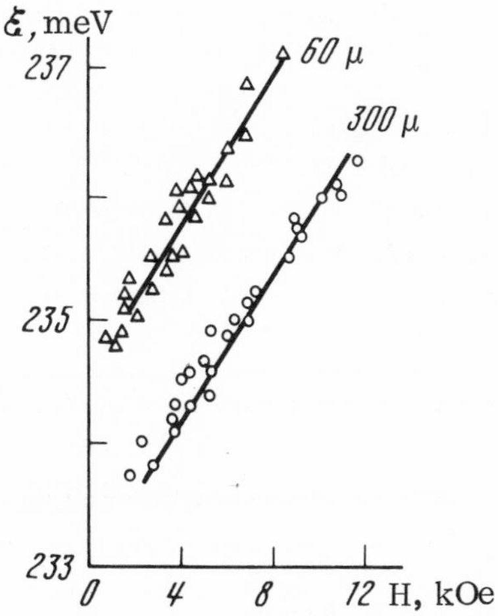

Fig. 25. Magnetic-field-dependences of the energy at the maximum representing the electron–light hole radiative transition in samples with $p_0 = 2 \cdot 10^{13}$ cm^{-3} and different thicknesses (lengths) of the p-type region, determined at T = 4.2°K.

with a slope $\Delta\mathcal{E}/\Delta H$ = 0.16 meV/kOe for the exciton level associated with the $m_s = +^1/_2$ sublevel and with a slope $\Delta\mathcal{E}/\Delta H$ = 0.16 meV/kOe for the $m_s = -^1/_2$ case. Equation (38) cannot be applied to the light-hole band because the effective mass of holes at the n = 0 Landau level should vary with the magnetic field (because of the complex structure of the valence band) and this should increase slowly the energy of the light holes at the n = 0 Landau level in a magnetic field, as demonstrated by Pidgeon and Brown [14]; therefore, no attempt was made to calculate the Landau level energy for the light holes in a magnetic field.

Thus, the application of a magnetic field increases the density of states at the Landau levels of electrons and light holes, which should increase the intensity of the recombination radiation associated with transitions between these levels. A longitudinal magnetic field does not interfere with the observation of the Franz–Keldysh effect [82], so that there should be the same change in the energy in a magnetic field in the case of the electron–light hole transitions in thick and thin samples. Within the limits of the experimental error, it was found that the change in the energy of the maximum corresponding to the electron–light hole transitions in a magnetic field was the same for the two types of sample, as demonstrated in Fig. 25. It is clear from this figure that the magnetic-field-induced change in the energy is the same for the long and short samples, in spite of the fact that the maximum for the long sample (L = 300 μ) observed in a magnetic field is shifted toward longer wavelengths because of the strong influence of the electric field. Measurements of the intensity and energy of the radiative transitions in magnetic fields indicated that this intensity passed through a maximum in the long and short samples and in both cases the maximum was located at 235 meV. This energy was close to the forbidden band width of crystals of this purity (see § 4 below) and, therefore, it was possible to assume that the energy was governed by the value of kT at the bottom of the valence band. The magnetic field in which this photon energy was reached in a long sample amounted to 8 kOe but for the short sample it was 1 kOe. The rate of change of the photon energy in the magnetic field was $\Delta\mathcal{E}/\Delta H$ = 0.31 meV/kOe, which did not correspond to any of the calculated values for the electron–heavy hole transitions (Fig. 24). The value of $\Delta\mathcal{E}/\Delta H$ = 0.31 meV/kOe, reduced by 0.26 meV/kOe for the Landau level of electrons with n = 0 and $m_s + ^1/_2$, showed that the rate of change of the light-hole Landau level energy in a magnetic field did not exceed $\Delta\mathcal{E}/\Delta H$ = 0.05 meV/kOe.

Thus, an investigation of the influence of electric and magnetic fields on the electron–light hole radiative transitions indicated that the electric field reduced the energy of these transitions and increased the light-hole temperature, whereas the magnetic field increased the density of states and, consequently, increased the intensity of the electron–light hole radiative transitions.

However, as shown above (Fig. 21), the application of a magnetic field may not only enhance but also reduce the intensity of the recombination radiation resulting from the electron–light hole transitions. The reduction occurs in high magnetic fields, at energies exceeding 235 meV, and it is due to an inverse redistribution of carriers between the light-and heavy-hole bands when the light-hole Landau level intersects either the Fermi level or an energy of the order of kT in the valence band.

We shall conclude by pointing out that in very weak magnetic fields of H = 1 kOe, the intensity of the electron–light hole radiative transitions falls strongly with the electric field but for a given value of the electric field the change in the intensity is governed by the density of states and energy of light holes in a magnetic field.

§ 4. Forbidden Band Width of Doped p-Type Indium Antimonide

Investigations of the recombination radiation emitted from germanium [83] and silicon [84] with different donor concentrations have demonstrated that an increase in the impurity center concentration reduces the energy of the interband radiative transitions but has hardly any effect on the energy of the transitions from a donor level to the valence band. An explanation of this behavior can be found in a paper by Volkov and Matveev [85], who calculated the changes in the energies of the main bands and obtained expressions valid in the impurity concentration range corresponding to $10 > \langle r\rangle/a > 2.5$, where $\langle r\rangle = (3/4\pi N_i)^{1/3}$; *a* is the Bohr radius of a carrier. They assumed that the carrier density was equal to the concentration of the impurity centers N_i and that T = 0. Volkov and Matveev demonstrated that the results of their calculations were in agreement with the experimental data reported in [84].

Thus, an increase in the concentration of the impurity centers reduces the energy of the photons corresponding to the interband radiative transitions [86]. However, if the effective electron and hole masses are equal, it is impossible to deduce the energy changes in the field of impurity centers from the recombination radiation because the influence of impurities on electrons and holes is of the same magnitude but of opposite sign.

Therefore, this effect is best observed in semiconductors characterized by very different electron and hole effective masses. The situation is particularly convenient in indium antimonide because of the very considerable difference between the effective masses of electrons and holes and also between the impurity-level energies so that a radiative transition can be observed at energies corresponding to an energy band edge. It is natural to use p-type indium antimonide for this purpose because the energy of the impurity centers in the p-type material

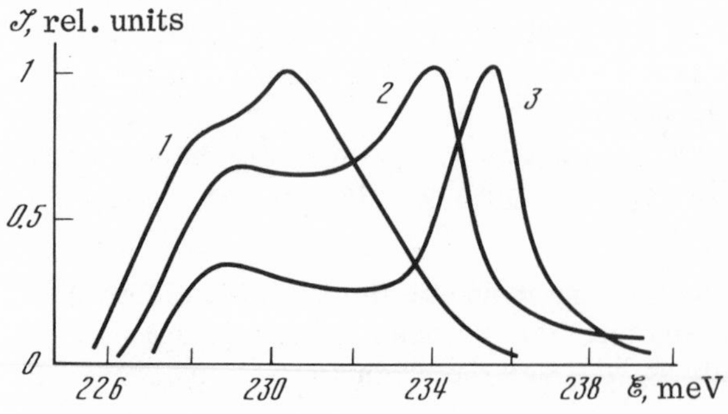

Fig. 26. Recombination radiation spectra of InSb obtained at T = 4.2°K at low rates of injection into samples with different acceptor concentrations: 1) $p_0 = 8 \cdot 10^{15}$ cm^{-3}; 2) $2 \cdot 10^{14}$ cm^{-3}; 3) $2 \cdot 10^{13}$ cm^{-3}.

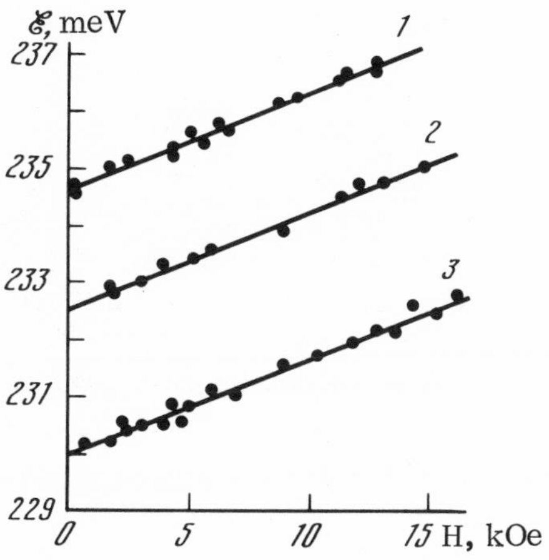

Fig. 27. Magnetic-field dependences of the energy at the maxima in the recombination radiation spectra recorded at T = 4.2°K for samples with different dopant concentrations: 1) $p_0 = 2 \cdot 10^{13}$ cm^{-3}; 2) $2.4 \cdot 10^{14}$ cm^{-3}; 3) $8 \cdot 10^{15}$ cm^{-3}.

is much higher than in the n-type material. Moreover, as shown above, the high density of states in the valence band makes it possible to observe recombination radiation at energies close to the band edge, when the injected-carrier density is still so low that electrons are distributed within a range of kT near the bottom of the band.

We determined the recombination radiation spectra of p-type indium antimonide samples with impurity center concentrations $p_0 = N_A - N_D = 2 \cdot 10^{13}$, $2 \cdot 10^{14}$, and $8 \cdot 10^{15}$ cm^{-3}.

Figure 26 shows the spectra obtained for these samples at injection levels corresponding to $n \approx p \approx N_i$ [87]. We can see that an increase in the impurity center concentration shifts the interband recombination radiation maximum toward longer wavelengths. A more accurate estimate of the shift can be obtained by measurements in a magnetic field H < 15 kOe. Figure 27 demonstrates the change in the energy corresponding to the electron–valence band radiative transitions resulting from the application of a magnetic field at injection rates such that $F_e \lessgtr kT$. We can see that the magnetic-field-induced change in the energy of the interband maximum corresponds to an exciton radiative transition. Thus, at low injection rates in magnetic fields $H \gtrless 2$ kOe there is only a radiative transition associated with an exciton level belonging to the Landau level characterized by n = 0 and $m_s = \pm{}^1/_2$. Since the energy of the exciton level is very low, $\mathscr{E}_{exc} \approx 2kT$ at T = 4.2°K, the exciton transition can be observed only if F_e is less than 2kT. The use of n-type indium antimonide and low injection rates enabled us to observe the exciton transition which was not revealed in the numerous early experiments carried out on n-type samples or at high injection rates.

Estimates of the shift of the interband transition energy with increasing impurity concentration, found by extrapolation to H = 0, gave the following values: there was no shift for a sample with an impurity concentration of $2 \cdot 10^{13}$ cm^{-3}; the shift was 2 meV when this concentration was $2 \cdot 10^{14}$ cm^{-3}; it increased to 4.7 meV for $p_0 = 8 \cdot 10^{15}$ cm^{-3}; these estimates were obtained assuming that the exciton level energy was 235 meV.

Since the range of the impurity center concentrations N_i from $2 \cdot 10^{13}$ to $8 \cdot 10^{15}$ cm^{-3} corresponds to the condition $\langle r \rangle / a > 10$, we can consider the pure Coulomb interaction between impurity centers without allowance for screening. In this case the change in the energy corresponding to the top of the valence band should be described by the expression $\Delta\mathscr{E} \simeq e^2/\mathscr{E}\langle r \rangle$. A calculation carried out using this relationship gives $\Delta\mathscr{E} = 4 \cdot 10^{-4}$ eV for $\Delta N_i = 2 \cdot 10^{13}$ cm^{-3}; $\Delta\mathscr{E} = 1 \cdot 10^{-3}$ eV for $\Delta N_i = 2 \cdot 10^{14}$ cm^{-3}; $\Delta\mathscr{E} = 3 \cdot 10^{-3}$ eV for $\Delta N_i = 8 \cdot 10^{15}$ cm^{-3}.

These values are in basic agreement with the experimental data but are somewhat smaller than the latter. Inclusion of corrections for the exchange interaction and correlation

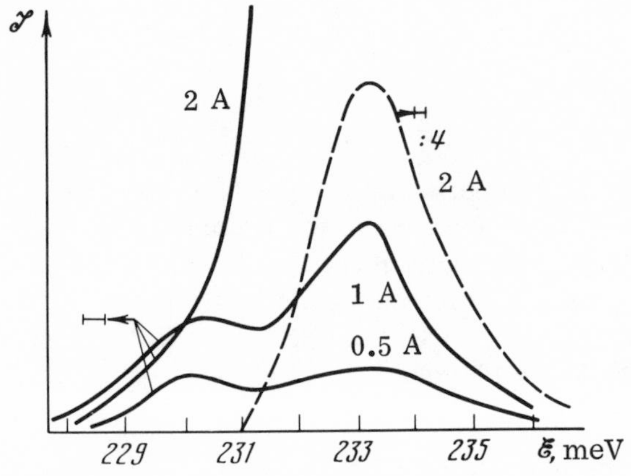

Fig. 28. Recombination radiation spectra of a sample with an uncompensated acceptor concentration $p_0 = N_A - N_D = 2 \cdot 10^{13}$ cm^{-3} recorded at T = 4.2°K for three currents (0.5, 1, and 2 A). The dashed curve for I = 2 A is the continuous curve replotted on a smaller scale.

improves the agreement between the experimental and theoretical data but it is not possible to estimate these corrections accurately for temperatures $T \neq 0$.

It is clear from Fig. 26 that for any carrier density the position of the energy of the maximum corresponding to the electron–acceptor level transitions is not affected, which indicates that the ground-state energy of the acceptor remains constant.

The energy and half-width of this transition remain unchanged right up to the injected-carrier density corresponding to $n = p = p_0$. At very high injection rates, when $n = p \gg p_0$, the intensities of the lines due to the transitions to the acceptor level become saturated and, which is more important, there is a reduction in the width of the recombination radiation band representing these transitions. This was manifested by a reduction in the intensity of the long-wave part of the spectrum of the electron–acceptor level radiative transitions, as shown in Fig. 28. Since the intensity of the electron–acceptor level radiative transitions is low, we can only speak of qualitative relationships. We shall assume that this behavior of the spectra is associated with the screening-induced reduction in the overlap of the wave functions of the impurity atoms at higher carrier densities.

CHAPTER IV

DOUBLE INJECTION OF CARRIERS AT 4.2°K INTO p^+-p-n^+ STRUCTURES BASED ON PURE p-TYPE InSb

Beginning from the first investigation of p^+-p-n^+ structures by Hall [88], much theoretical and experimental work has been done on conduction in such structures [89, 90]. Most of this work has been concentrated in the range of low currents, i.e., in the negative resistance region, but there have been several investigations carried out at high rates of injection into these structures [91]. Lampert [90] carried out the most thorough calculations of the current–voltage (I–V) characteristics of these structures for a range of impurity concentrations, temperatures, and rates of injection. Therefore, we shall use Lampert's results in discussing the experimental data.

§ 1. Low Injection Rates

It follows from Lampert's theory [90] that at low injection levels the current is dominated by the majority carriers, in accordance with Ohm's law

$$j = ep_0\mu_p V/L, \tag{43}$$

where j is the current density; μ_p is the hole mobility; V is the voltage applied only to the p-type region; L is the length of the p-type region; p_0 is the majority-carrier density.

When the minority-carrier lifetime (in our case the electron lifetime) becomes comparable with the carrier drift time, Ohm's law is replaced by a quadratic dependence, called by Lampert the "semiconductor regime."

This occurs at a voltage V_Ω satisfying the condition .

$$V_\Omega = L^2/\mu_e\tau_e = L^2\sigma_e^0 v_T N_R^0/\mu_e, \tag{44}$$

where μ_e is the electron mobility; v_T is the thermal velocity of electrons; N_R^0 is the concentration of neutral recombination centers; σ_e^0 is the electron-capture cross section of a neutral recombination center. In the voltage range $V > V_\Omega$ the current through the sample should be governed by the dependence

$$j = \frac{9}{8}\varepsilon\,[\mu_e V^2/L^3], \tag{45}$$

where ε is the permittivity.

The dependence

$$j \propto V^2$$

is observed until the majority-carrier lifetime becomes equal to the transit time. A negative resistance should begin at a turnover voltage $V_{to} = L^2/\mu_p\tau_p$, which corresponds to the equality of the majority-carrier lifetime and the transit (drift) time. The voltage then falls with rising current and the minimum voltage is given by [90]:

$$V_{min} = (\sigma_e^0/\sigma_p^-)\,V_{to}, \tag{46}$$

where σ_p^- is the hole-capture cross section of an ionization center.

Samples with the p-type region L > 150 μ long were observed by us to have a negative resistance, typical of transition from conduction due to one type of carrier to conduction due

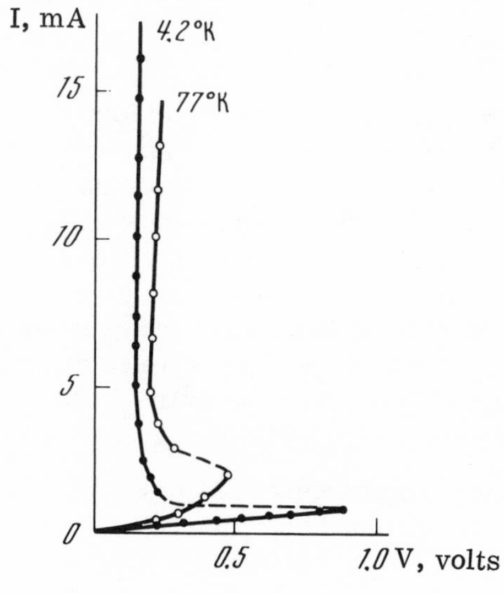

Fig. 29. Current–voltage characteristics of a sample with L = 300 μ recorded at 4.2 and 77°K for low injection rates.

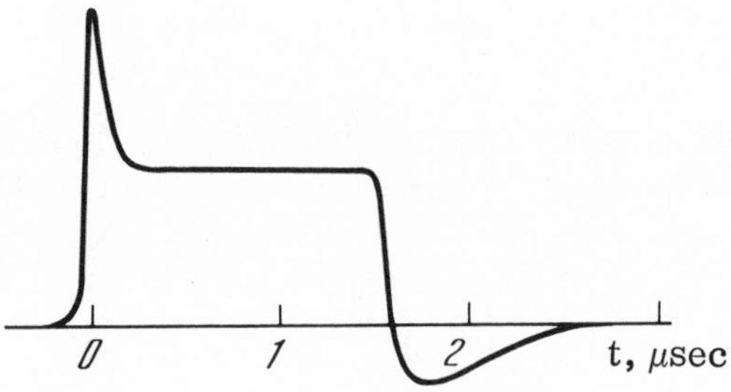

Fig. 30. Shape of a voltage pulse.

to two types of carrier. Figure 29 shows the current-voltage (I−V) characteristics of a sample with a p-type region L = 300 μ long subjected to low injection rates at T = 4.2 and 77°K. A negative resistance is observed at both temperatures. At 77°K the law corresponding to the semiconductor regime $j \propto V^2$ can be observed quite clearly but at 4.2°K it is hardly noticeable and the turnover takes place at a voltage $V_{to} \leq V_{\Omega}$. This shows that $\mu_e \tau_e \approx \mu_p \tau_p$ for pure p-type InSb crystals at 4.2°K. The formation of a negative resistance region is accompanied by the appearance of a recombination radiation line due to electron−acceptor level transitions. An analysis of the shape of the current and light pulses can be used to estimate the carrier lifetime τ_1 at a current corresponding to V_{to}. Since the shape of the light pulses does not repeat accurately the shape of the current pulses, it follows that the carrier lifetime, at the injection rate employed, is of the order of $5 \cdot 10^{-7}$ sec. The ambipolar drift time of carriers across the sample can be estimated from the time taken to establish steady-state conditions and this can be done from the shape of the voltage pulse exactly as in [92]. Figure 30 shows the shape of the voltage pulses applied to the sample with the p-type region L = 300 μ long. This shape of the voltage pulses is due to variation in the resistance of the sample. Electrons are injected initially from the n^+−p contact (cathode) and then the space charge is compensated by an equal number of holes arriving through the opposite p^+−p contact. At the beginning of a pulse the major part of the crystal is not yet filled with carriers and it has a considerable resistance, which governs the initial voltage. However, as a plasma cloud spreads across the crystal to the p^+-type contact, the resistance and voltage fall to steady-state values, which correspond to complete filling of the crystal with this plasma. The voltage decay time can be used to estimate the plasma drift time τ_{dr}. It amounts to $2 \cdot 10^{-7}$ sec. Thus, we can say that the condition $\tau_1 \approx \tau_{dr}$ necessary for the observation of a negative resistance when the crystal is filled with the plasma is satisfied at currents of 10^{-3} A and voltages of 1-2 V (depending on the length of the crystal).

§ 2. High Injection Rates

After the turnover point the conditions in the p-type region correspond to high injection levels (the current beyond this point is high). The current−voltage characteristics can be analyzed on the basis of Lampert's theory for insulators where it is assumed that: 1) the intrinsic carrier density can be ignored; 2) the densities of the injected electrons and holes are approximately equal, i.e., $n \approx p$; 3) the current is limited by recombination via traps; 4) the average carrier density in the middle of the insulator is governed approximately by the neutral plasma (this assumption gives rise to an error of 1%); 5) the lifetimes of two types of carrier are approximately equal and depend on the carrier density; 6) the transit times of carriers of both kinds are much shorter than the lifetime; 7) the space charge due to the difference between the densities of carriers of two signs does not affect the mobile charges.

When these assumptions are made, one can expect dependences of the $j \propto V^3$ type and, at high injection rates, also dependences of the $j \propto AV^2$ type and such that the current varies but V = const [93].

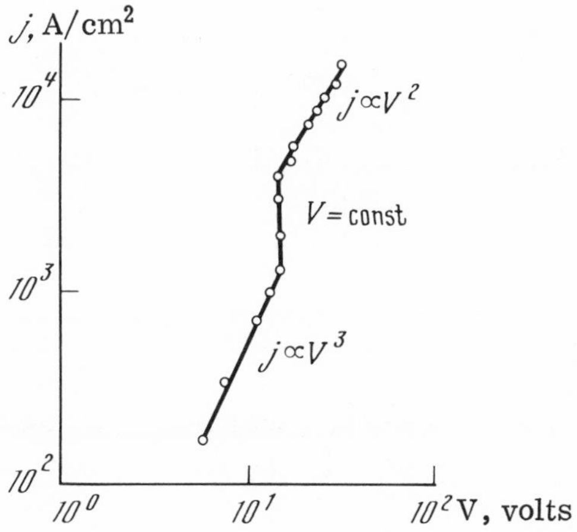

Fig. 31. Current–voltage characteristic of a sample with L = 750 μ at T = 4.2°K.

Samples with the p-type region L = 750 μ long were found to exhibit all three laws with rising rates of injection. Samples with L ⩽ 450 μ exhibited only the dependences $j \propto V^2$ and $j \propto V$ and the second one appeared at higher injection rates than the first. Figure 31 shows a j–V characteristic obtained for a sample with L = 750 μ. We can see that as the rate of injection is increased, a long sample exhibits all three regions of the j–V characteristic predicted by Lampert's theory for long structures ($L_a < L$). Here, L_a is the ambipolar diffusion length defined by

$$L_a = \sqrt{2\frac{kT}{e}\mu_e\tau/(1+b)}\ , \qquad (47)$$

where $b = \mu_e/\mu_p$ and τ is the carrier lifetime.

According to Kleinman [94], the $j \propto V^2$ law can be observed at high injection rates in samples with $L \ll L_a$ and $L \gg L_a$, but then the dependences of j on L are quite different. If $L \ll L_a$, then $j \propto L^2$, but for $L \gg L_a$, we have $j \propto \exp(L/L_a)$.

The nature of the injection conditions was determined; i.e., it was found whether the rate of injection was high or low, by preparing samples with different lengths of the p-type region and using the dependences of j on L at a given voltage to find whether the dependence $j \propto V^2$ indicated a low or a high injection rate.

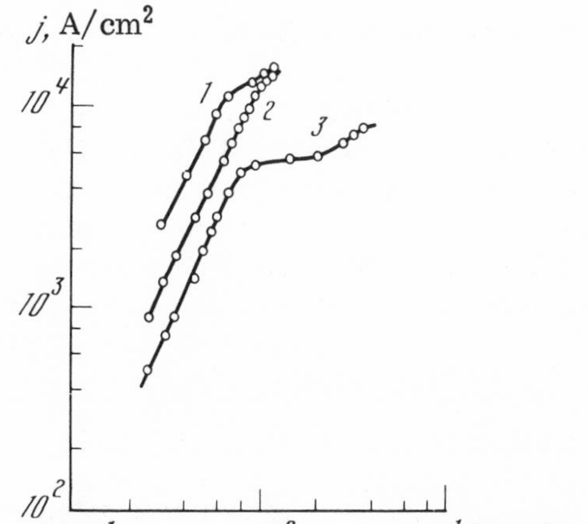

Fig. 32. Dependences of the current density on the voltage recorded at T = 4.2°K for three samples prepared from p-type InSb with an impurity concentration of $2 \cdot 10^{13}$ cm^{-3} and different lengths of the p-type region: 1) 200 μ; 2) 300 μ; 3) 450 μ.

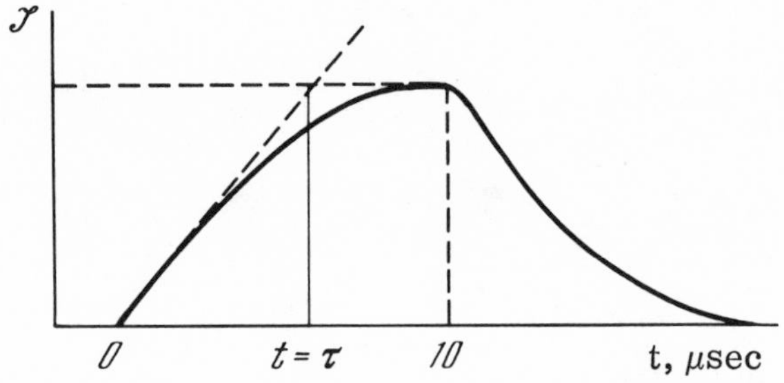

Fig. 33. Schematic illustration of the method for determining the lifetime from the shape of the recombination radiation pulses.

Figure 32 shows the dependences of j on V for three samples with p-type regions 450, 300, and 200 μ long. The dependence of j on L at a given value of V can be seen to be close to the dependence $j \propto L^2$.

The carrier lifetimes were estimated from the shape of the light pulses, as illustrated schematically in Fig. 33. Such estimates gave 4-6 μsec for the lifetime in the region of the quadratic dependence of the current on the voltage, where the electron–light hole radiative transitions predominated. At all currents corresponding to the dependence $j \propto V^2$ the lifetime estimated above remained constant.

A comparison of the j–V characteristics for $L \leq 450\ \mu$ and $L = 750\ \mu$ indicated that in the case of the samples with $L = 450\ \mu$ the condition $L/L_a < 1$ was satisfied, whereas for the samples with $L = 750\ \mu$, we found that $L = 750\ \mu$, $L/L_a > 1$. Hence, we concluded that $450\ \mu < L_a < 750\ \mu$. Assuming that $L_a = 500\ \mu$, $kT = 4 \cdot 10^{-4}$ eV, and $\tau = 5 \cdot 10^{-6}$ sec, we found that $\mu_e(1 + b)^{-1} = L_a^2 e/2kT\tau = 8 \cdot 10^5\ \mathrm{cm^2 \cdot V^{-1} \cdot sec^{-1}}$, i.e., $\mu_p \gtrsim 4 \cdot 10^5\ \mathrm{cm^2 \cdot V^{-1} \cdot sec^{-1}}$.

However, it is known that at T = 4.2°K the hole mobility does not exceed $10^4\ \mathrm{cm^2 \cdot V^{-1} \cdot sec^{-1}}$ even in the purest indium antimonide crystals. This value is governed by the scattering on ionized impurities. The maximum value of the mobility in the purest p-type indium antimonide crystals is observed at T ≈ 20-30°K [35]. Below this temperature the mobility decreases because of the stronger scattering by ionized impurities. The temperature dependences of the mobility in the range T ≥ 30°K are described satisfactorily by the dependence $\mu \propto T^{-3/2}$, which is typical of the scattering by the acoustic vibrations of the lattice. If the dependence $\mu \propto T^{-3/2}$ is extrapolated to low temperatures, it is found that at T = 4.2°K the mobility in the purest crystals may reach $10^6\ \mathrm{cm^2 \cdot V^{-1} \cdot sec^{-1}}$.

Thus, a mobility of $4 \cdot 10^5\ \mathrm{cm^2 \cdot V^{-1} \cdot sec^{-1}}$ may be exhibited by holes only if we assume that the scattering by the acoustic vibrations of the lattice predominates at T = 4.2°K. However, as shown in Chap. I § 3, the light carriers should dominate conduction in the case of scattering by acoustic vibrations. Therefore, we cannot ignore the contribution of the light holes to the electrical conductivity. It is difficult to calculate this contribution because of the complex structure of the valence band. It is shown by Pikus [95] that when the interband transitions between the light- and heavy-hole subbands have a high probability and when the light and heavy holes are transferred from one to the other in a distance equal to several mean free paths, the electrical conductivity can be expressed in the form $\sigma_p = \sigma_1 + \sigma_2$:

$$\sigma_p = \sigma_2\left(1 + \frac{\sigma_1}{\sigma_2}\right) = e\mu_2 p_2\left(1 + \frac{\mu_1 p_1}{\mu_2 p_2}\right), \tag{48}$$

$$\mu_p = \mu_2\left(1 + \frac{\mu_1 p_1}{\mu_2 p_2}\right) \Big/ \left(1 + \frac{p_1}{p_2}\right). \tag{49}$$

Using Eq. (48), we can show that the heavy holes dominate the electrical conductivity in the case of scattering by impurities and that the light holes dominate the conductivity when the scattering is by lattice vibrations. However, according to Bir, Normantas, and Pikus [96], interband transitions between the light- and heavy-hole subbands can make a considerable contribution to the scattering in crystals with a complex valence band, similar to that found in germanium. This interband scattering is particularly important if the relaxation times in the subbands are governed by the scattering on acoustic vibrations. The calculation reported in [96] applies to the elastic scattering by acoustic phonons in the case of heavy- and light-hole bands of spherical symmetry. These calculations show that the relaxation times in the light- and heavy-hole subbands become equal and there is a considerable deviation from the equilibrium distribution function in the light-hole subband.

The quantitative results for this effect are inapplicable to holes in the valence band of indium antimonide at T = 4.2°K because of the considerable anisotropy of the heavy-hole subband ($m_{\parallel} : m_{\perp} = 3$) and also because the scattering of the heavy holes by acoustic vibrations cannot be regarded as elastic. In fact, at T = 4.2°K for $m_1 = 0.5m_0$ the wave vector of the phonons interacting most effecctively with holes satisfies the condition $\mathbf{q} = \mathbf{k} = (2kTm_1/\hbar^2)^{1/2}$ and it is so large that the phonon energy $\hbar qS$ (S is the velocity of sound) is of the order of the carrier energy.

As soon as allowance for the anisotropy (even a weak anisotropy, such as that found in germanium) reduces strongly the influence of the interband transitions on the mobility in each of the subbands, we can assume – in the first approximation – that the contribution of the light holes can be estimated from the formula $\sigma = \sigma_1 + \sigma_2$ assuming that the mobility in each of the subbands is described by the usual formulas without allowance for the interband transitions although the effect considered by Bir and Pikus [97] cannot be excluded completely.

Thus, in the region of the quadratic dependence of the current on the voltage the electrical conductivity and recombination radiation include a considerable contribution from the light holes and this results in a long mean free path of carriers and in a constant lifetime in the case of the interband radiative transitions.

In the case of shorter (thinner) samples with $L \lesssim 150\ \mu$, we found that the dependence $j \propto V$ was observed in a wide range of currents and it followed $j \propto V^2$. The transition to the linear dependence and high current densities has been observed by many investigators but it has not been explained satisfactorily; many authors assume that, in this range of currents, it is essential to allow for the dependence of the lifetimes on the injected-carrier densities [92]. In fact, the transition to bimolecular recombination, not considered by Lampert, can give rise to a less steep dependence of j on V, as found by Barnett [93] for the $j \propto V^3$ regime when allowance was made for the dependence of the lifetime on the injected-carrier density.

Figure 34 shows the j–V characteristic of a sample with L = 150 μ. The current density was calculated from the area of the lower contact, which was much larger than the upper

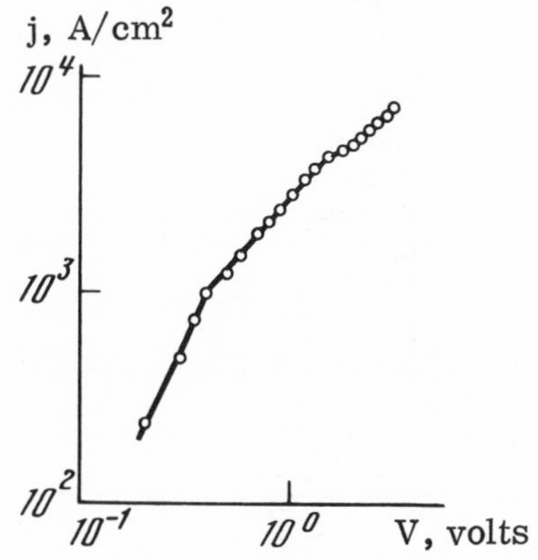

Fig. 34. Dependence of j on V recorded at T = 4.2°K for a sample with a p-type region 150 μ long and different upper and lower contact areas.

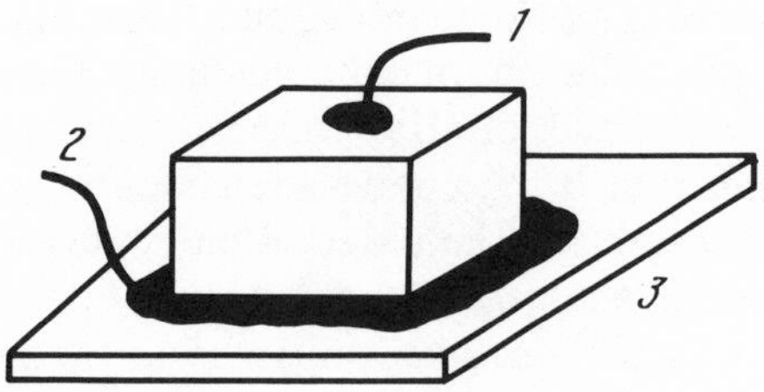

Fig. 35. Schematic diagram of a diode with a smaller upper-contact area: 1) upper contact; 2) lower contact; 3) copper support.

contact. This upper contact had an indefinite shape and its area was considerably smaller than the surface of the sample, as shown in Fig. 35. It is clear from Fig. 35 that in the range $j < 10^3$ A/cm^2 (low current densities) the current depends quadratically on the voltage, but in the range $j > 10^3$ A/cm^2 the dependence is linear. This transition to the linear dependence of j on V is manifested by an increase in the intensity of the electron–heavy hole radiative transitions in the recombination radiation spectra and by a reduction in the radiative lifetime with increasing rate of injection.

§ 3. Influence of Magnetic Fields on I – V Characteristics

The influence of magnetic fields on the I–V characteristics is particularly important at high injection levels. In the region of the quadratic dependence of the current on the voltage an increase in the magnetic field results in a fall and then in a rise of the resistance of the sample, and a negative resistance region is observed in magnetic fields corresponding to a maximum of the intensity of the electron–light hole radiative transitions. Investigations of the Shubnikov–de Haas effect in lightly doped p-type indium antimonide crystals also revealed a negative magnetoresistance [38] in a magnetic field in which the light-hole Landau level crossed the Fermi level or the energy kT in the valence band. A study of the Shubnikov–de Haas effect in heavily doped p-type indium antimonide crystals at T = 4.2°K demonstrated that a magnetic field produced special features in the magnetoresistance oscillations which were associated with a strong rise of the probability of the transfer of a hole from the heavy to the light band.

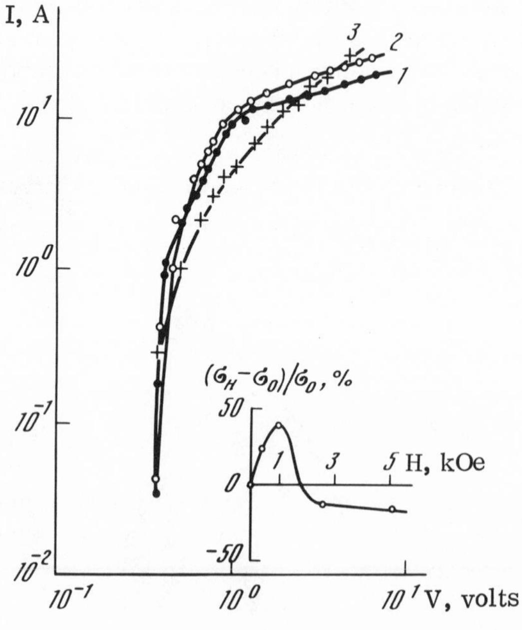

Fig. 36. Current–voltage (I–V) characteristics obtained at 4.2°K using three values of the magnetic field: 1) H = 0; 2) 500 Oe; 3) 5000 Oe. The inset shows the dependence of the electrical conductivity on the magnetic field for I = 8 A.

It follows from these investigations that the application of a magnetic field causing quantization of the light-hole band into Landau levels can alter the probability of the interband transitions so strongly that w_{12} becomes considerably greater than w_{21} (w_{21} and w_{12} are the probabilities of the transfer of a hole from the light to the heavy band and vice versa). This considerable change in the probabilities of transitions between the light- and heavy-hole bands in both directions may affect the hole mobility and the equilibrium densities of the light and heavy holes in the valence band. The behavior of the recombination radiation spectra (see Chap. III, § 2) shows that the change in the probability of the interband transitions in a magnetic field increases the relative density of the light holes in the valence band. Therefore, the change in the resistance in a magnetic field may be attributed to a change in the density of the light holes participating in electrical conduction although some contribution to the change in the resistance may also be due to a change in the mobility.

Figure 36 shows the current–voltage (I–V) characteristics of a sample with $L = 250\ \mu$ obtained in various magnetic fields. It is clear from this figure that in the range of currents where the conductivity is a quadratic function of the current in $H = 0$, there is a maximum of the conductivity in a magnetic field $H \sim 1$ kOe (inset). Strong magnetic fields reduce the electrical conductivity and, in the region where the current depends quadratically on the voltage, the conductivity in a magnetic field $H > 1.5$ kOe is less than in the absence of the field. This increase in the resistance in a magnetic field, compared with the resistance in the absence of the field, is observed only in the region where the current is a quadratic function of the voltage. The reduction in the electrical conductivity in a longitudinal magnetic field compared with the zero-field conductivity may be due to a reduction in the density of the light holes or due to the nonparabolicity of the valence band of the heavy holes. However, the fall of the intensity of the recombination radiation resulting from the electron–light hole transitions suggests a reduction in the light-hole density in strong magnetic fields in which the light-hole Landau level is displaced beyond the Fermi level in the valence band [73].

Figure 37 shows how a current at a given voltage varies with the magnetic field in different parts of the I–V characteristics of a sample with $L = 270\ \mu$. It is clear from this figure that the quantity $(I_H - I_0)/I_0$ becomes negative only in the range of currents where $j \propto V^2$ but when the current is higher or lower, there is only a rise of the current in the magnetic field and approach to $(I_H - I_0)/I_0|_{H\to\infty} \to 0$ in strong magnetic fields. This behavior of the I–V characteristics can be explained satisfactorily by an increase in the probability of the interband transitions from the heavy- to the light-hole band and by an increase in the light-hole density in weak magnetic fields. One should also mention that the smallest rise of the current is observed in the region of the quadratic dependence of j on V, where the contribution of the light holes to the electrical conductivity is considerable even for $H = 0$.

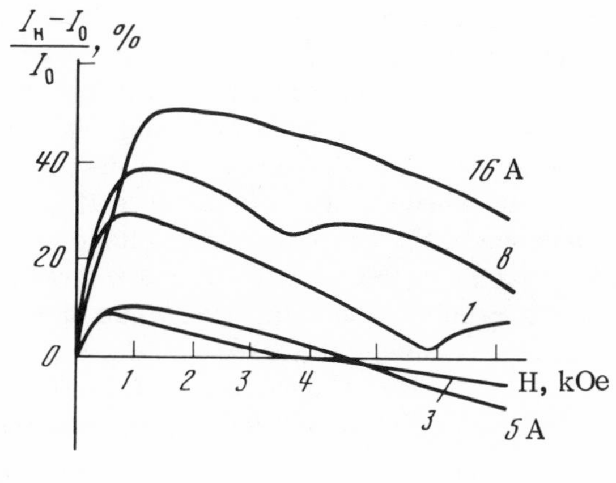

Fig. 37. Dependences of $(I_H - I_0)/I_0$ on the magnetic field for a sample with a p-type region 250 μ long obtained for different currents given alongside each figure.

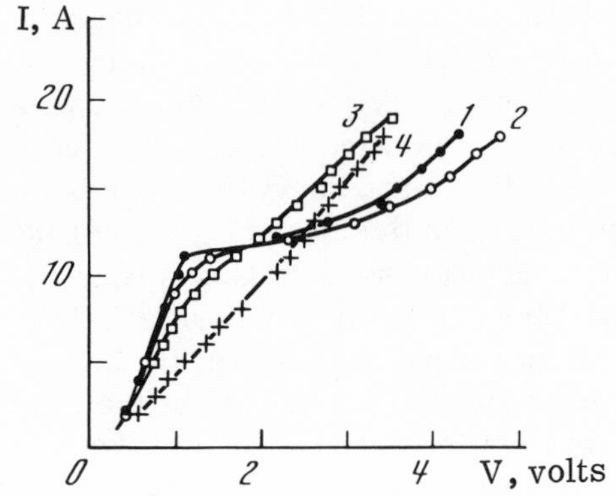

Fig. 38. Current–voltage (I–V) characteristics of a sample with L = 150 μ obtained in different magnetic fields: 1) H = 0; 2) 500 Oe; 3) 2 kOe; 4) 5 kOe.

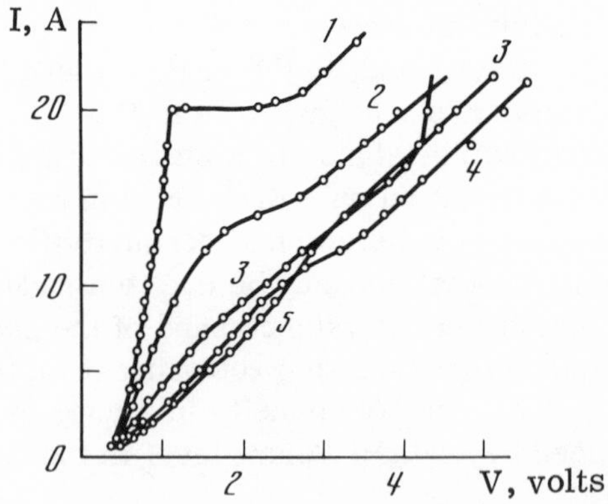

Fig. 39. Current–voltage (I–V) characteristics of a sample with L = 230 μ in a transverse magnetic field H ⊥ E: 1) H = 0; 2) 25 Oe; 3) 50 Oe; 4) 74 Oe; 5) 125 Oe.

The behavior of the I–V characteristics in the region of the linear dependence of the current on the voltage cannot be described in a simple manner. One may observe either a negative value of the quantity $(I_H - I_0)/I_0$, or a positive value (Fig. 38). Figure 38 shows the I–V characteristics of a sample with L = 150 μ recorded in various magnetic fields. We can see that in the range $I \gtrsim 8$ A, where the current varies linearly with the voltage, there is no negative resistance. This range of currents has not been considered theoretically and the reason for this linear dependence is not clear, so it is difficult to explain the results obtained in magnetic fields.

All the results reported above were obtained in longitudinal magnetic fields. However, it should be mentioned that studies of the Shubnikov–de Haas effect revealed a negative magnetoresistance in longitudinal and transverse fields. However, under our conditions there was no negative magnetoresistance in transverse fields. Figure 39 shows the I–V characteristics of a sample with L = 230 μ subjected to transverse magnetic fields. We can see that the resistance of the sample increases with the magnetic field and when this field is close to 125 Oe, the resistance reaches saturation. The quadratic dependence of the current on the voltage is observed in the range I < 10 A. We can consider the behavior of the quantity $(I_H - I_0)/I_0$ with increasing transverse magnetic field in this range of currents. The saturation value of the magnetic field can be used to estimate the mobility on the assumption that $j \propto \mu_p V^2$. It is then found that $\mu_p H/c \approx 3$ for $\mu_p = 4 \cdot 10^5$ cm$^2 \cdot$V$^{-1} \cdot$sec^{-1}. This is equal to the value of $4 \cdot 10^5$ cm$^2 \cdot$V$^{-1} \cdot$sec^{-1} obtained for the mobility from the ambipolar diffusion length in H = 0 (see § 2).

CHAPTER V

EMISSION OF COHERENT RADIATION FROM INDIUM ANTIMONIDE AT 4.2 AND 20°K

As shown above (see Chap. I, § 5), coherent radiation can be generated if the condition $F_e - F_p > h\nu$ is satisfied, where $h\nu$ is the photon energy, and F_e, F_p are the quasi-Fermi levels of electrons and holes. In the case of a pure semiconductor under double injection conditions, when $n = p \gg n_0, p_0$, the quasi-Fermi levels of electrons and holes can be estimated ignoring the presence of impurity centers. At a finite temperature the quasi-Fermi level energies can be expressed in terms of their values at absolute zero using a formula suggested by McDougall and Stoner [98]:

$$\left[\frac{F(0) - \mathscr{E}_{c,v}}{kT}\right]^{3/2} = \frac{3\pi^{1/2}}{4}\mathscr{F}_{1/2}(\eta).$$

Here, $\mathscr{F}_{1/2}(\eta) = 2\pi^{1/2}\int\limits_0^\infty \frac{\mathscr{E}^{1/2}d\mathscr{E}}{1+\exp\left(\frac{\mathscr{E}}{kT} - \eta\right)}$ is the Fermi–Dirac integral and $F(0) = \mathscr{E}_{c,v} \pm \frac{h^2}{2m_{c,v}}\left(\frac{3n_0, p_0}{8\pi}\right)^{2/3}$ is the Fermi energy of a carrier at T = 0.

The Fermi–Dirac integrals have been tabulated for a wide range of η. The calculation of F(0) presents no difficulties. Figure 40 gives the carrier-density dependences of the reduced $F(T)/kT = \eta$ quasi-Fermi levels of electrons and holes at temperatures of 4.2 and 20°K; it also gives the carrier density dependences of the difference $\eta_e - \eta_p$ at the same temperatures. The minimum injected-carrier densities necessary to achieve $\eta_e - \eta_p > h\nu$ can be estimated from these curves because the minimum energy of the photons produced by the interband transitions corresponds to $\eta_e - \eta_p = 0$. We can see that although in a sample with $n = 2 \cdot 10^{14}$ cm^{-3} the quasi-Fermi level of electrons is located inside the conduction band at T = 4.2°K, the population inversion condition for the interband transitions is satisfied only when the injected-carrier densities are $n = p \approx 10^{14}$ cm^{-3}. However, a population inversion is only a necessary but not a sufficient condition for the generation of coherent radiation. We must also satisfy a more difficult condition

$$K = \beta + \frac{\ln(R_1R_2)^{-1}}{2L}. \tag{50}$$

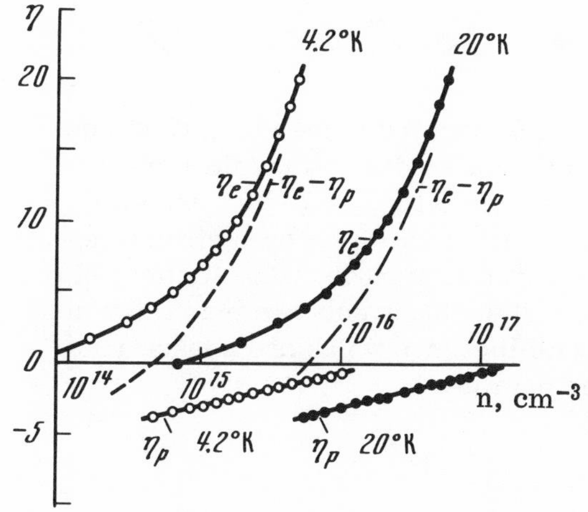

Fig. 40. Calculated [98] dependences of the 4.2 and 20°K quasi-Fermi levels of electrons and holes on the carrier density and the corresponding dependences of the difference $\eta_e - \eta_p$ at the same temperatures. The positive values of η are measured from the edges into the allowed bands and the negative values lie in the forbidden band. The difference $\eta_e - \eta_p$ is calculated under the standard conditions, i.e., when the middle of the forbidden band is taken as zero.

Here, K is the gain; β is the loss factor; L is the resonator length; R_1 and R_2 are the reflection coefficients of the resonator faces.

The gain K for the direct interband transitions is governed by the difference between the absorption and emission at a given photon energy. For a specific transition and a given photon energy, K is a function of the distribution of carriers in the conduction and valence (light- and heavy-hole) bands, i.e., [58]:

$$K = A \mid M \mid^2 \sqrt{\mathscr{E}} \, [f_e(\mathscr{E}) - f_v(\mathscr{E}_c - h\nu)]. \tag{51}$$

The population inversion conditions for the two radiative transitions 1 and 2 can be different in spite of the same rate of injection because the energy at the maximum in the spectrum of the electron–light hole radiative transitions lies in the range $\mathscr{E} < \mathscr{E}_g + F_e$, whereas for electron–heavy hole transitions it is $\mathscr{E} = \mathscr{E}_g + F_e$. Hence, it follows that the gain for the electron–light hole transitions (1) is greater than the gain for the electron–heavy hole transitions (2) even when the rate of injection is the same. Since the wavelengths of the electron–light hole and electron–heavy hole radiative transitions differ only slightly, we may assume that the radiative losses are the same in both cases. However, the gain for the transition (1) should be considerably greater than the gain for the transition (2) also because of the slower rise of the negative absorption coefficient with increasing energy because of the strong anisotropy of the heavy-hole band.

For all these reasons the threshold densities of the current needed to generate coherent radiation as a result of the electron–light hole transitions should be lower than those for the electron–heavy hole transitions.

§ 1. Emission of Coherent Radiation at 4.2°K

Figure 41 shows the dependences of the energies of the coherent radiation modes on the rate of injection of carriers into a sample whose p-type region is L = 80 μ long. It is clear from this figure that if the current is I < 5 A, coherent radiation modes appear at photon energies corresponding to the electron–light hole transitions. An increase in the current increases the intensities of these modes. When the current becomes I = 10 A, modes corresponding to the electron–heavy hole transitions are observed. In the range I > 14 A, the intensity

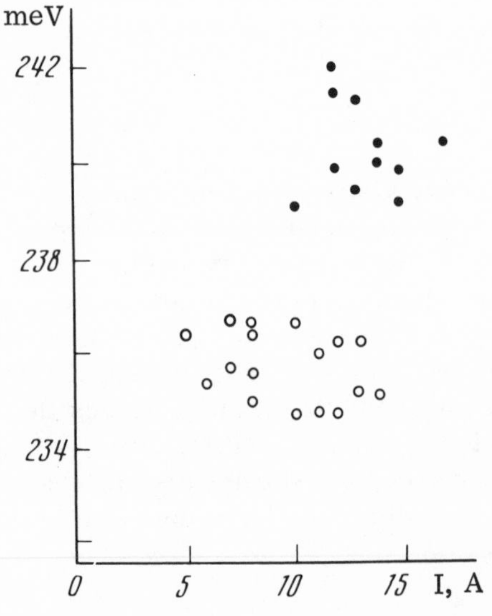

Fig. 41. Dependences of the energies of coherent radiation modes on the injection current in a sample with $p_0 = 2 \cdot 10^{13}$ cm^{-3} and L = 80 μ at T = 4.2°K. The open circles represent the radiation modes with lifetimes of 4.6 μsec and the black dots are the radiation modes with lifetimes which are a function of the injection rate.

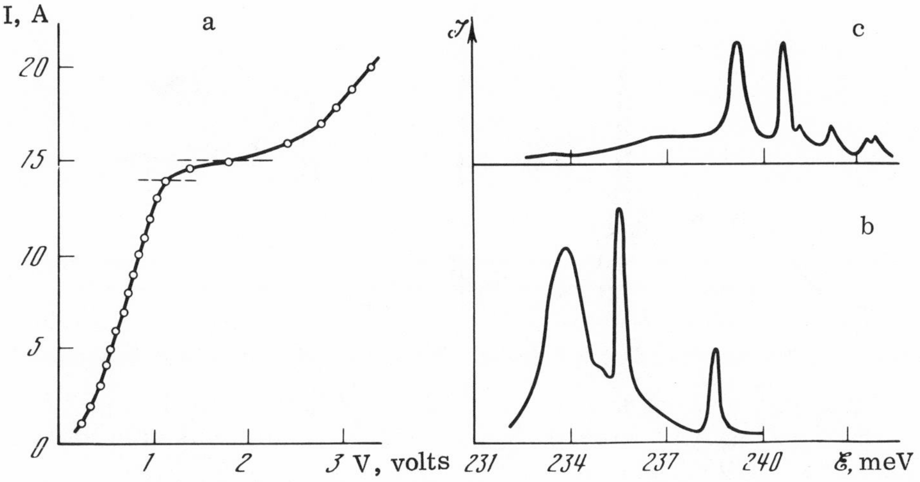

Fig. 42. Current–voltage (I–V) characteristic (a) and coherent emission spectra of a sample with L = 300 μ obtained using a current I = 14 A (b) and I = 15 A (c).

of the former modes decreases and the intensity of the latter rises. The appearance of the line due to the electron–heavy hole transitions is accompanied by a fall of the total radiation intensity. The decrease in the intensities of the lines due to the two radiative transitions cannot be attributed to overheating because the modes generated as a result of the electron–heavy hole transitions have higher photon energies, which suggests that $F_e - F_p > h\nu$.

Figure 42 shows the emission spectrum obtained at I = 14 and 15 A as well as the current voltage (I–V) characteristic. We can see that when the current is I = 14 A, the emission spectrum is dominated by the electron–light hole transitions whereas for I = 15 A the electron–heavy hole transitions are the most important. This change in the radiative transition mechanism is accompanied by a fall of the total radiation intensity and a simultaneous decrease in the electrical conductivity of a sample because of a strong reduction in the carrier lifetime.

The threshold current density for the generation of coherent radiation as a result of the transition 1 depends very strongly on the electric field. Figure 43 shows the dependence of this threshold current density on the average electric field in a sample for currents near the threshold value. The dependence is clearly due to the influence of the electric field on the light holes.

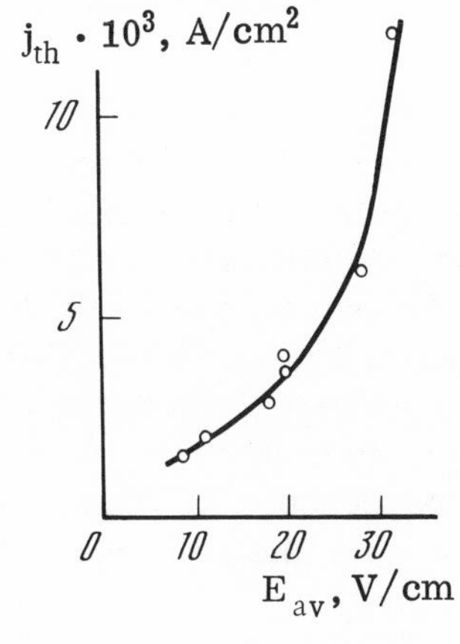

Fig. 43. Dependence of the threshold current on the average electric field at T = 4.2°K.

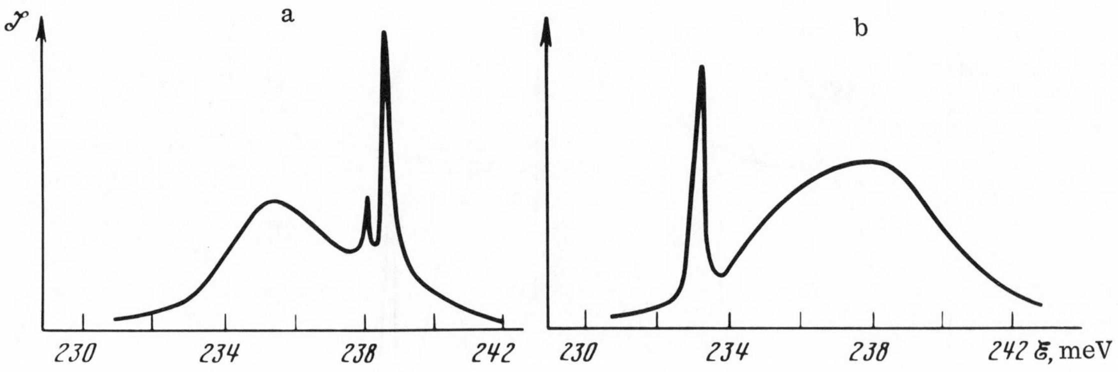

Fig. 44. Stimulated emission spectra obtained at T = 4.2°K at the threshold for two samples with the same impurity concentration of $2 \cdot 10^{13}$ cm^{-3} but with different lengths of the p-type region: a) 450 μ; b) 150 μ.

Figure 44 gives the emission spectra of two samples with p-type regions 150 and 450 μ long. The value of the energy at the maximum of the spontaneous emission spectrum of the sample with L = 150 μ and the energy of the coherent radiation modes obtained near the threshold from the sample with L = 450 μ can be used to estimate the quasi-Fermi level of electrons in both samples. It is found that F_e = 4-5 meV, i.e., that $n = p = 2 \cdot 10^{15}$ cm^{-3} in both samples. However, the maxima in the spontaneous emission spectra of these two samples occur at quite different energies. In the case of the long sample with L = 450 μ the spontaneous emission maximum is shifted toward longer wavelengths and the coherent radiation mode appears at higher energies corresponding to the electron–heavy hole transitions. In the case of the thinner sample the energy of the coherent emission line corresponds to the electron–light hole transitions and the maximum of the recombination radiation spectrum is due to the electron–heavy hole transitions. It is shown in Chap. III, § 3 that the energy of the electron–light hole radiative transitions depends strongly on the electric field. An increase of this field shifts the energy of these transitions toward longer wavelengths and reduces the intensity of the corresponding band or line. The fall of the intensity is due to the heating of the light holes by the electric field. It is necessary to point out that in the same field the relative heating of electrons [defined as $T = T_{eff}(0) - T_{eff}(E)$, where $T_{eff}(0) \sim F_e$] is weak because at higher rates of injection the Fermi level for the electron gas is $F_e \gg kT$ (at T = 4.2°K). These electric fields do not alter either the temperature of the heavy holes, which is indicated by the appearance of the coherent radiation at photon energies corresponding to the electron–heavy hole transitions when the current corresponds to the injected-carrier density $n = p = 5 \cdot 10^{15}$ cm^{-3}.

§ 2. Dependence of the Threshold Current on the Magnetic Field

A strong magnetic field quantizes the conduction and valence bands into Landau levels in a plane perpendicular to the direction of the electric field applied at the same time. This occurs in sufficiently strong magnetic fields which satisfy the conditions $\omega\tau > 1$ and $\hbar\omega_c > kT$. Our study of pure indium antimonide crystals revealed a strong dependence of the threshold current on the magnetic field (Fig. 45). A magnetic field first caused a steep fall of the threshold current from 20 to 8 A and then – after a minimum at 8 kOe – there was a small rise of the threshold current. A study of the influence of the magnetic field on the energies of the coherent emission lines indicated that the magnetic-field dependence of the threshold current was governed by a fall of the threshold current for the electron–light hole transitions and a rise of the threshold current for the electron–heavy hole transitions.

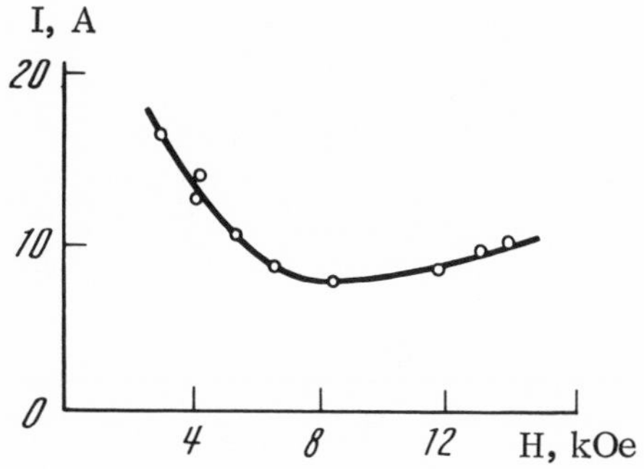

Fig. 45. Dependence of the threshold current on the magnetic field applied to a sample with $p_0 = 2 \cdot 10^{13}$ cm^{-3} and L = 300 μ at T = 4.2°K.

Figure 46 demonstrates the dependences of the energies of two coherent emission lines obtained for a sample whose magnetic-field dependence of the threshold current is plotted in Fig. 45. It is clear from Figs. 45 and 46 that in the range of magnetic fields corresponding to a strong fall of the threshold current only one coherent emission line is emitted and its energy varies with the magnetic field at a rate $\Delta\mathscr{E}/\Delta H = 0.31$ meV/kOe and this line is due to the electron–light hole transitions. In a magnetic field H = 6 kOe a short-wavelength emission line appears and its energy is ~236 meV. The energy of this line varies with the magnetic field at a rate of $\Delta\mathscr{E}/\Delta H = 0.16$ meV/kOe. An increase in the magnetic field causes the intensity of this electron–heavy hole line to increase. In magnetic fields H > 10 kOe, the electron–light hole line is no longer observed.

Coherent radiation emitted in the presence or absence of a magnetic field is polarized and the polarization vector of the radiation emitted as a result of the electron–light hole as well as for the electron–heavy hole transitions is perpendicular to the magnetic field, i.e., $\mathbf{E} \perp \mathbf{H}$. As demonstrated by Pidgeon and Brown [14], the intensity of the electron–light hole transitions may be six times as high as the intensity of the electron–heavy hole transitions producing radiation of the same polarization.

Figure 47 shows the coherent emission spectra obtained for a sample with L = 300 μ in magnetic fields of H = 6, 8, and 12 kOe. We can see how the main transition changes in a field H = 8 kOe. The mode spacing (deduced from the coherent emission spectrum in a field H = 8 kOe) can be used to estimate the refractive index from

$$\Delta\lambda = \lambda^2/2L\left(n - \lambda\frac{dn}{d\lambda}\right).$$

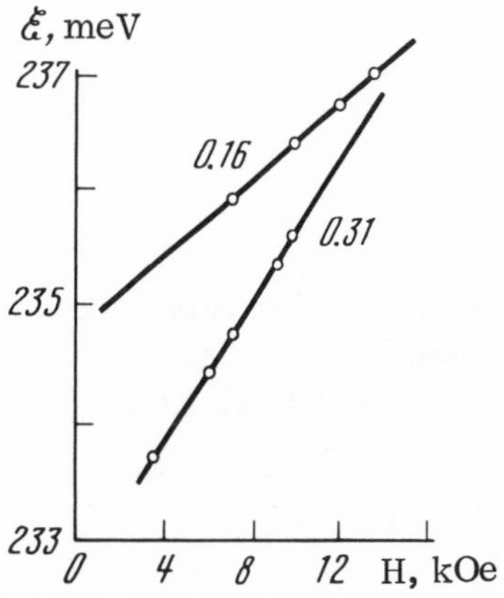

Fig. 46. Dependences of the energies of two coherent radiation modes on the magnetic field applied to a sample with $p_0 = 2 \cdot 10^{13}$ cm^{-3} and L = 300 μ when the current was I = 18 A and temperature was T = 4.2°K.

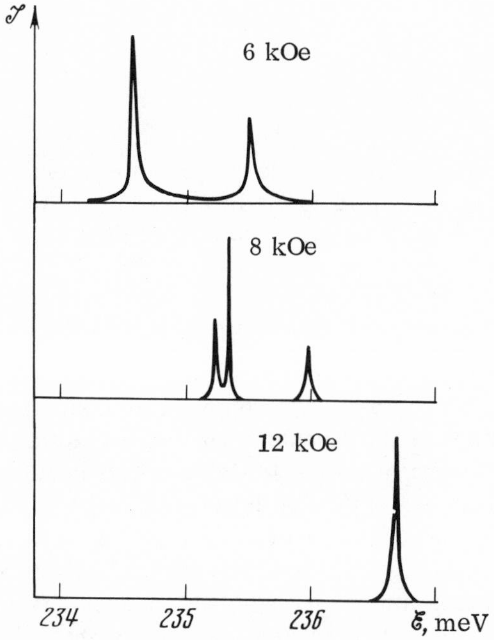

Fig. 47. Coherent emission spectra of sample with $p_0 = 2 \cdot 10^{13}$ cm^{-3} and L = 300 μ at T = 4.2°K, obtained for three values of the magnetic field.

If Δλ = 40 A and L = 650 μ (distance between the reflecting planes), we find that $\bar{n} = n - \lambda\, dn/d\lambda = 5.2$, which is in good agreement with the value obtained in [24].

As pointed out earlier (see § 1 in the present chapter), at high rates of injection the electron–light hole transitions are practically negligible. Under these conditions the electron–heavy hole transitions predominate. At high rates of injection in a magnetic field we can observe two radiative transitions from the $m_s = \pm 1/2$ sublevels, belonging to the n = 0 Landau level, to the heavy-hole band. In the absence of a magnetic field a radiation mode appears at $\mathscr{E} = 237.3$ meV, as shown in Fig. 48 (these results were obtained for a sample with L = 300 μ). In magnetic fields H ≤ 2 kOe the conduction band becomes quantized into the Landau levels and this increases the long-wavelength part of the emission spectrum (Fig. 49). In a field H = 3 kOe, corresponding to $g\beta H = 2.5$ kT (T = 4.2°K), there are two radiative transitions associated

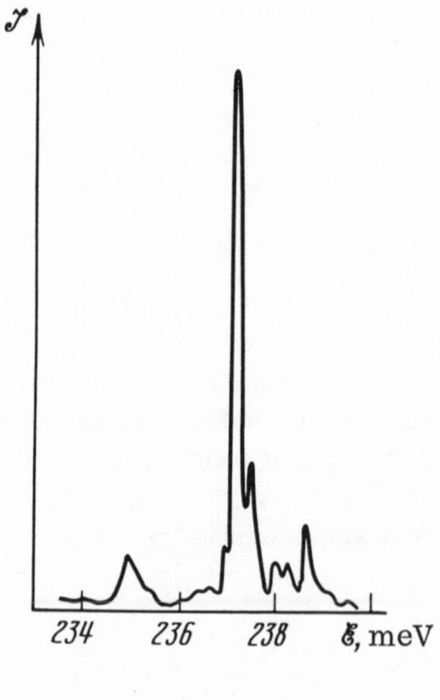

Fig. 48. Coherent emission spectrum of a sample with L = 300 μ and $p_0 = 10^{15}$ cm^{-3} at T = 4.2°K for I = 18 A.

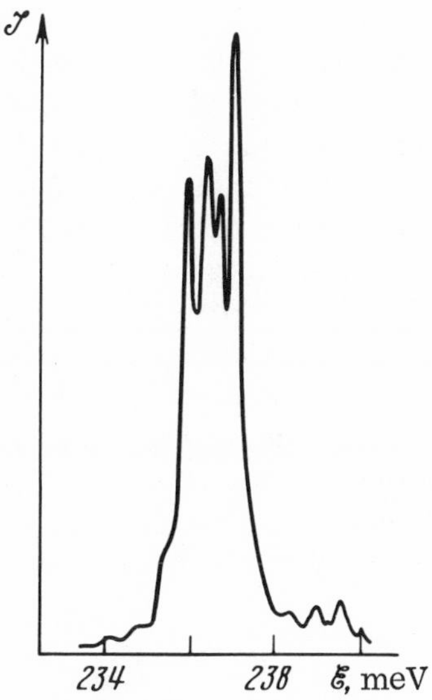

Fig. 49. Coherent emission spectrum of a sample with $p_0 = 10^{15}$ cm^{-3} and L = 250 μ obtained in a field H = 1.5 kOe at a temperature T = 4.2°K for a current I = 18 A.

with the n = 0 Landau level and corresponding to the $m_s = \pm 1/2$ sublevels. An increase in the magnetic field enhances the intensity of the line due to transitions of electrons from the n = 0, $m_s = -1/2$ level to the heavy-hole valence band. This increase in the intensity is observed right up to magnetic fields H = 6 kOe. The energy of the coherent radiation modes in a magnetic field H = 6 kOe, in which the intensity of the radiation due to this transition has a maximum, is approximately equal to the Fermi level energy if the latter is estimated from the energy of the coherent emission line in H = 0 ($\mathscr{E}$ = 237.6 meV).

Magnetic fields in the range H > 6 kOe redistribute electrons so that they fill the Landau level with n = 0 and $m_s = +1/2$. This should result from the crossing of the Landau level with n = 0 and $m_s = -1/2$ and the electron Fermi level. The fall of the intensity of the transitions from the $m_s = -1/2$ level and the rise of the intensity of the transitions from the $m_s = +1/2$ level occur right up to magnetic fields H = 8 kOe and when H = 8 kOe is reached practically all the electrons are localized in the Landau level with n = 0 and $m_s = +1/2$. Since the width of the Landau level does not exceed kT, it follows that magnetic fields H > 8 kOe lift the degeneracy and the exciton state becomes favored by the energy considerations. An electron and

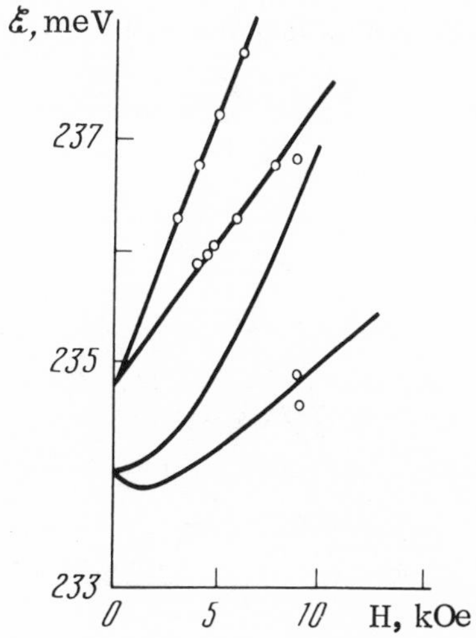

Fig. 50. Dependences of the energies of coherent emission lines on the magnetic field applied to a sample with L = 250 μ at T = 4.2°K. The continuous curves are the calculated dependences and the points are the experimental values.

a heavy hole form an exciton and when a magnetic field H = 9 kOe is applied, coherent radiation is observed at photon energies corresponding to an exciton transition with $\mathcal{E}_{max} \approx 234.5$ meV. Figure 50 shows the calculated magnetic-field dependences of the energy of the Landau level with n = 0 and $m_s = +1/2$ and also of the exciton level associated with this Landau level. The calculation was made assuming $m_s = 0.014 m_0$ for the effective electron mass, g = −50 for the g factor of electrons, and $\mathcal{E}_{exc} = 0.7$ meV for the binding energy of the exciton level in H = 0. The dependences were calculated by the method of Yafet, Keyes, and Adams [81]. We can see from Fig. 50 that the agreement between the calculated dependences and experimental points was good. Extrapolation of the Landau level to zero field gave the forbidden band width (for a given value of the current). This width was $\mathcal{E}_g = 234.8$ meV. The Fermi energy could be estimated from the difference between the forbidden band width and the energy corresponding to the maximum intensity of the emission line resulting from the transitions of electrons from the level with n = 0 and $m_s = +1/2$ to the valence band. This gave $\Delta\mathcal{E} = -\mathcal{E}_g + \mathcal{E}_{max} = 237.6 - 234.8 = 2.8$ meV, which corresponded to an injected-carrier density $n > 10^{15}$ cm^{-3}.

§ 3. Emission of Coherent Radiation at 20°K

An increase in the lattice temperature alters the temperature of carriers in the conduction and valence bands and this lifts the degeneracy and reduces the value of $\eta_e - \eta_p$ at a given carrier density. It is clear from Fig. 40 that the strong increase of the threshold current can be expected for pure InSb crystals because at 20°K the condition $\eta_e - \eta_p = 0$ is satisfied at injected carrier densities $\approx 7 \cdot 10^{15}$ cm^{-3}, i.e., at densities which are an order of magnitude higher than at 4.2°K. If the same degree of degeneracy is required for coherent radiation emission at 20°K as at 4.2°K, the carrier density should be an order of magnitude higher because

$$\frac{\eta\,(20^\circ\,\mathrm{K})}{\eta\,(4.2^\circ\,\mathrm{K})} = \left[\frac{n\,(20^\circ\,\mathrm{K})}{n\,(4.2^\circ\,\mathrm{K})}\right]^{3/2} = \mathrm{const},$$

if

$$\frac{n\,(20^\circ\,\mathrm{K})}{n\,(4.2^\circ\,\mathrm{K})} = 10.$$

If the minimum threshold current corresponds, as before, to the electron–light hole transitions, a change in the threshold carrier density by an order of magnitude implies that there is no change in the radiation losses.

As shown below, the emission wavelength at 20°K differs little from the wavelength at 4.2°K. This means that the reflection losses depend weakly on temperature because $\bar{n}(4.2°K) =$

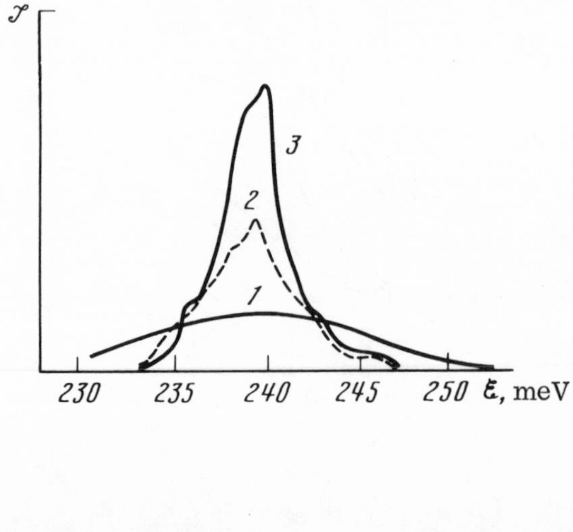

Fig. 51. Recombination radiation spectra of pure indium antimonide obtained for three values of the injection current at T = 20°K: 1) I = 8 A; 2) I = 12 A; 3) I = 17 A.

5.2 (see § 2), compared with $\bar{n} = 5$ at 20°K. The free-carrier absorption can also contribute to the radiation losses in the $h\nu > F_e - F_p$ case. In indium antimonide the free-carrier absorption is mainly due to interband transitions between the light- and heavy-hole valence subbands. As for any other interband mechanism, the probability of such transitions is governed by their matrix element, densities of states, and carrier distribution functions in the relevant bands. A slight change in the photon energy because of the temperature rise from 4.2 to 20°K allows us to assume that the matrix element is unaffected by the temperature. Therefore, the change in the reabsorption by free carriers is governed only by the changes in the carrier distribution functions of the light- and heavy-hole bands at energies corresponding to $\mathbf{k} = 10^{-2}$ cm^{-1} when the gap between the two subbands is $\Delta\mathscr{E} = 2 \cdot 10^{-1}$ eV [see Fig. 1 and Eqs. (2) and (3)]. At 4.2 and 20°K the probability of the hole occupancy of the levels with energies corresponding to the wave vector $\mathbf{k} = 2 \cdot 10^{-1}$ cm^{-1} is practically zero for the light- and heavy-hole subbands so that we can ignore the reabsorption of the radiation by free carriers at these temperatures.

Thus, it follows from our discussion that the loss factor can be regarded as constant between 4.2 and 20°K. This means that the emission of stimulated radiation at 20°K should occur at injected-carrier densities an order of magnitude higher than at 4.2°K.

Figures 51 and 52 show the spontaneous and coherent emission spectra obtained at 20°K for an indium antimonide sample with $p_0 = N_A - N_D = 6.3 \cdot 10^{13}$ cm^{-3} using various currents.

It is clear that an increase in the current reduces the width of the spontaneous radiation band, which is due to the appearance of the stimulated emission at $I \gtrsim 12$ A. The narrowing of the band occurs at energies close to the maximum in the spontaneous emission spectrum. Coherent radiation modes appear at 20 A and when the current reaches I = 25.5 A, the intensities of these modes exceed considerably the intensity of the spontaneous radiation band. The mode energy is close to the energy corresponding to the spontaneous radiation maximum, i.e., it is close to the Fermi energy in the case of the electron–heavy hole transitions or close to $\mathscr{E} \approx 2kT$ in the case of the electron–light hole transitions.

The temperature shift of the energy band edge of indium antimonide is $\mathscr{E}_g(T) = 2.7 \cdot 10^{-4}$ eV/deg. Therefore, when the temperature of the sample increases from 4.2 to 20°K, the forbidden band width of indium antimonide should decrease by

$$\Delta\mathscr{E}_g = 2.7 \cdot 10^{-4} \times 15.8 = 4.3 \text{ meV}.$$

For the same degree of carrier degeneracy the energy of the spontaneous radiation maximum is separated by $\mathscr{E} - F_e \approx 14$ meV from the new energy band edge. Thus, the shift of the energy of this maximum in the direction of shorter wavelengths should be even greater than that found

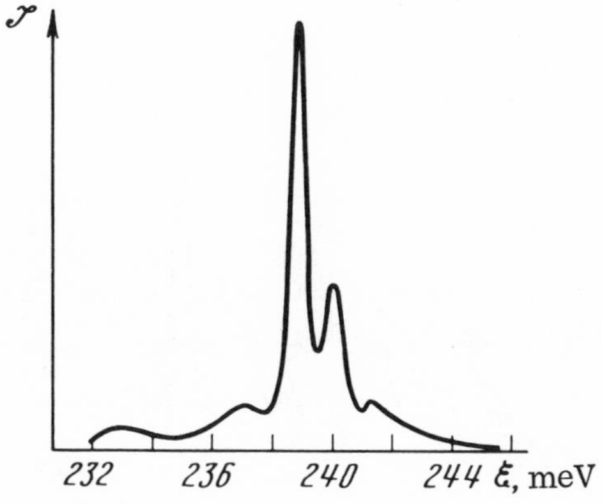

Fig. 52. Coherent emission spectrum of a sample with $p_0 = 6.3 \cdot 10^{13}$ cm^{-3} obtained at 20°K for I = 25.5 A.

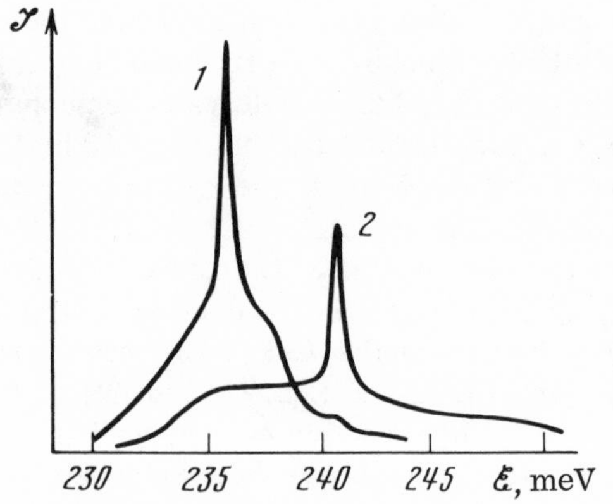

Fig. 53. Coherent emission spectra of a sample with $p_0 = 6 \cdot 10^{13}$ cm^{-3}: 1) I = 4 A, T = 4.2°K; 2) I = 20 A, T = 20°K.

experimentally. Figure 53 shows the coherent radiation spectra obtained for a sample with $p_0 = 6 \cdot 10^{13}$ cm^{-3} at 4.2 and 20°K. It is clear from this figure that the difference between the mode energies at the threshold is ≈6 meV but the current increases only by a factor of 5 between 4.2 and 20°K. The threshold current for this sample at 4.2°K is higher than the minimum threshold current and the energy at the maximum is fairly high, which correspondings to the electron–heavy hole transitions at 4.2°K. At 20°K the energy of the coherent radiation mode at the threshold suggests either the electron–heavy hole transitions or a change in the forbidden band width between 4.2 and 20°K by an amount smaller than that given by Eq. (52).

Coherent emission spectra obtained at currents much higher than the threshold value are shown in Fig. 54. We can see that the emission spectrum is now of the multimode type. The spacing between the modes can be used to estimate the refractive index at 20°K. If L = 450 μ and $\Delta\lambda$ = 60 Å, we find that $\bar{n} = n - \lambda dn/d\lambda = 5.0$. This is slightly less than at 4.2°K (see § 1) but the change is slight and it cannot influence greatly the threshold current.

§ 4. Emission of Coherent Radiation from Doped and Compensated InSb at 4.2 and 20°K

Figure 55 shows schematically the spontaneous emission spectra of pure InSb crystals obtained using a fairly high rate of injection at temperatures of 4.2, 20, and 77°K. The shaded parts identify regions of appearance of coherent radiation modes at the lowest current densities. We can see that as the temperature of the pure InSb crystal increases, the coherent radiation energy shifts toward shorter wavelengths because of an increase in the threshold current.

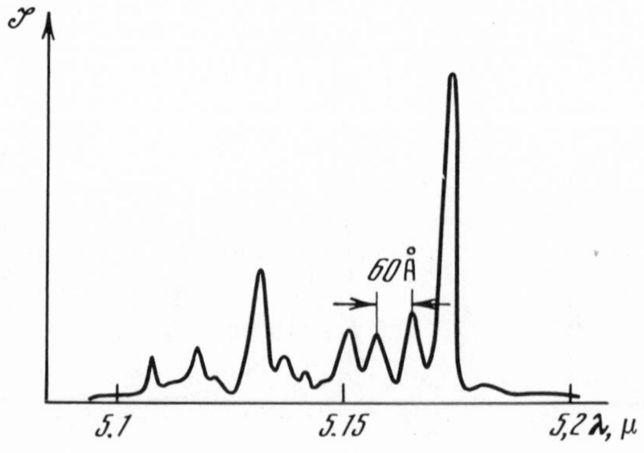

Fig. 54. Coherent emission spectrum of a sample with $p_0 = 2 \cdot 10^{13}$ cm^{-3} obtained at 20°K for a current of 19 A, well above the threshold.

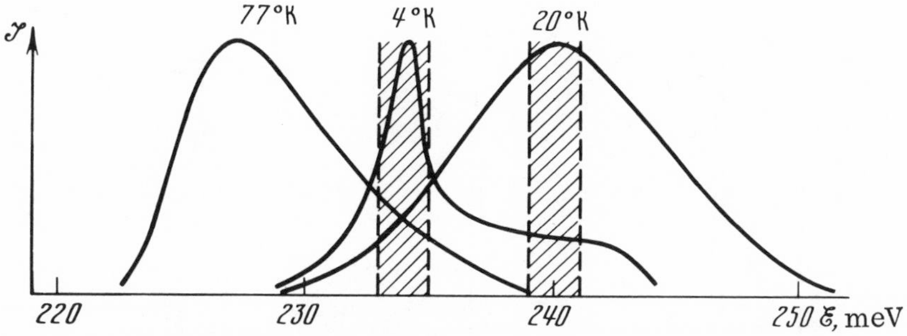

Fig. 55. Typical spontaneous and coherent (shaded) emission spectra of pure indium antimonide crystals recorded at 4.2, 20, and 77°K.

A study of compensated and doped p-type indium antimonide revealed that at 4.2°K the threshold current density for the samples with $N_A - N_D = 10^{16}$ cm^{-3} was 4-6 times higher than the value for pure indium antimonide with $N_A - N_D = 10^{14}$ cm^{-3}. There was also a reduction in the rate of rise of the threshold current between 4.2 and 20°K. The absolute value of the threshold current at 20°K increased slowly with the impurity concentration up to $N_A + N_D \approx 10^{16}$ cm^{-3}. This behavior may be due to a characteristic feature of the quasi-Fermi levels of electrons and holes of the doped and compesated samples. A calculation of the quasi-Fermi level at 20°K for $\mathcal{E}_I$ = 7-10 meV is very difficult even for the doped samples (the difficulties are even greater for the compensated material) and, therefore, we can speak only of a qualitative change in the behavior of the quasi-Fermi level. Introduction of excess acceptors produces additional electron traps so that the free-electron density is smaller by $n = N_A - N_D$ than the free-hole density. This corresponds to a reduction by $\eta_e' = F_e'/kT = (\hbar^2/2m_e kT) \times$

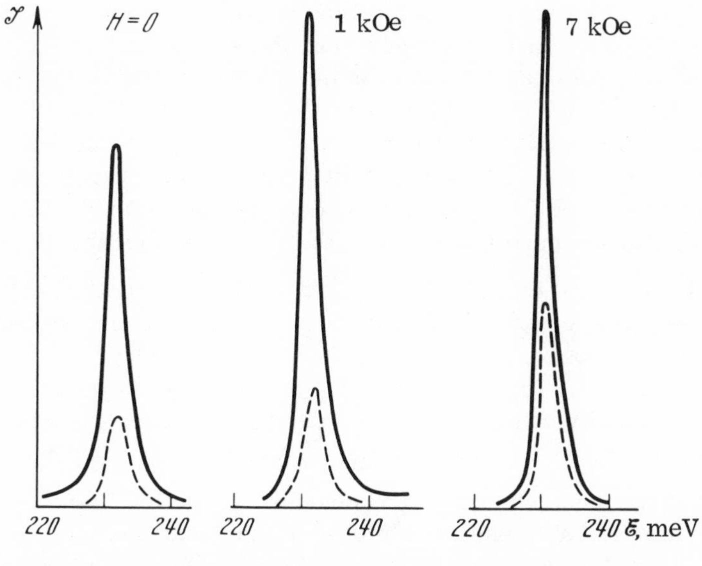

Fig. 56. Emission spectra of a compensated sample (K = 0.97) with $p_0 = 1.3 \cdot 10^{15}$ cm^{-3} recorded in different magnetic fields at different temperatures for I = 18 A. The continuous curves are the spectra recorded at 4.2°K and the dashed curves those recorded at 20°K.

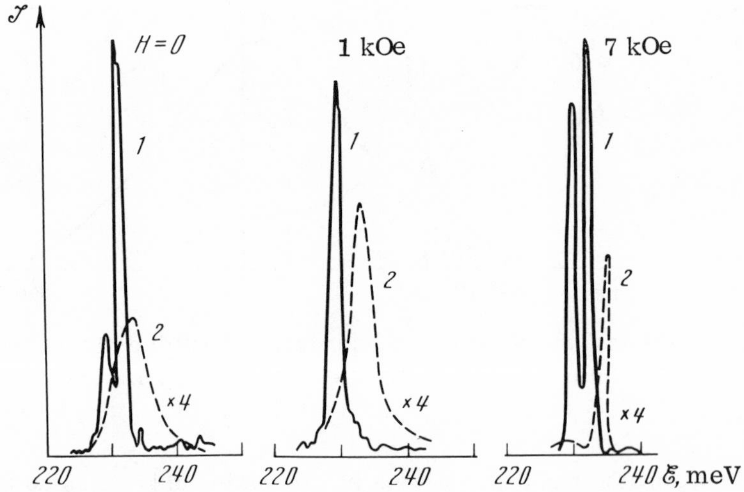

Fig. 57. Emission spectra of a p-type sample with $p_0 = 3 \cdot 10^{15}$ cm^{-3} obtained for I = 16 A using different magnetic fields at different temperatures: 1) 4.2°K; 2) 20°K.

$[3(N_A - N_D)/8\pi]^{2/3}$ in η_e for any injection level and it shifts the whole $\eta_e - \eta_p$ curve toward higher carrier densities, relative to the corresponding curve for the pure material. The shift is considerable for the doped and compensated materials, and this shift is governed by the difference $N_A - N_D$. This explanation should apply right up to impurity concentrations corresponding to the valence band degeneracy.

Figures 56 and 57 show the recombination radiation spectra of samples with impurity concentrations $N_A + N_D = 4 \cdot 10^{16}$ cm^{-3} (Fig. 57) and $N_A + N_D = 5 \cdot 10^{15}$ cm^{-3}, recorded at 4.2 and 20°K in different magnetic fields. A comparsion of the spectra obtained at 4.2°K in H = 0 shows that an increase in the impurity concentration shifts the emission line toward longer wavelengths compared with the lines obtained for the pure material (Fig. 56). The shift increases with the impurity concentration in the sample. This behavior can be expected in the case of a strong reduction in the forbidden band width as a result of an increase in the impurity concentration (see Chap. III, § 4). The energy shift toward longer wavelengths is considerable because even a strong rise of the injected-carrier density does not shift the energy back to the values obtained at 4.2°K for the pure materials at the minimum threshold current [87].

The difference between the energies at the line maxima and between the threshold currents at 4.2 and 20°K is exhibited only by samples with impurity concentrations in the range $N_A + N_D \gtrless 5 \cdot 10^{15}$ cm^{-3} but not by samples with $N_A + N_D \approx 4 \cdot 10^{16}$ cm^{-3} and this is true both in H = 0 and in magnetic fields. The reason for this behavior is not clear but we may suppose that an increase in the impurity concentration reduces the influence of temperature on the emission spectra in a manner similar to that observed for n-type indium antimonide [47].

Figures 56 and 57 demonstrate also a considerable reduction in the threshold current at 20°K as a result of application of a magnetic field to a doped sample with an impurity concentration $N_A + N_D = 5 \cdot 10^{15}$ cm^{-3}, as well as a long-wavelength shift of the lines observed at 4.2 and 20°K when the magnetic field is increased. However, the range of magnetic fields is limited (H $\lessdot$ 15 kOe) and, moreover, the intensity of the radiation emitted by the doped samples at 20°K is considerably less than the intensity of the radiation emitted by the pure samples at 4.2°K, so that the cause of the reduction in the threshold current and of the long-wavelength shift of the coherent radiation maximum in magnetic fields cannot yet be identified.

CONCLUSIONS

1. Electron–light hole radiative transitions were observed for the first time in interband recombination radiation emitted from indium antimonide. Coherent emission as a result of electron–light hole and electron–heavy hole transitions was obtained. The minimum threshold current for the generation of coherent radiation corresponded to the electron–light hole transitions.

2. At high rates of injection of carriers into pure indium antimonide crystals a considerable role was played by acoustic lattice vibrations even at T = 4.2°K and the light holes made a considerable contribution to the electrical conductivity.

3. A new effect, which was a redistribution of carriers between the light- and heavy-hole valence subbands, was observed in a magnetic field. This was observed when the magnetic field was sufficient for the quantization of the light-hole subband into Landau levels but not for the quantization of the heavy-hole subband. The redistribution of carriers was due to a change of the density of states and of the Landau level energy in the light-hole subband when a magnetic field was applied.

4. A negative magnetoresistance was observed in relatively weak magnetic fields. A simultaneous study of the current–voltage (I–V) characteristics and recombination radiation spectra established that the negative magnetoresistance was due to an increase (because of a redistribution between the subbands) of the proportion of the light holes participating in electrical conduction.

5. A magnetic field first reduced strongly the threshold current needed for the generation of coherent radiation at T = 4.2°K and then increased slightly this current. This behavior of the threshold current was associated with the redistribution of carriers between the light- and heavy-hole valence subbands and was correlated with the change in the light-hole density in magnetic fields.

6. Tuning of the emission frequency of the laser radiation at rates from 0.16 to 0.55 meV/kOe was achieved in magnetic fields using various transitions (exciton radiative transition, electron–light hole transitions, and electron–heavy hole transitions).

7. Coherent radiation was generated at T = 20°K in pure and doped p-type indium antimonide crystals. It was established that at T = 20°K the threshold current density depended weakly on the dopant concentration right up to $6 \cdot 10^{15}$ cm^{-3}.

8. An increase in the acceptor concentration in the p-type material resulted in a reduction in the forbidden band width, due to the Coulomb interaction between free carriers and charged centers. This phenomenon was used to vary the laser emission frequency from 235 meV for pure samples to 230 meV in the case of heavily doped crystals.

LITERATURE CITED

1. M. I. Nathan, W. P. Dumke, G. Burns, F. H. Dill, Jr., and G. Lasher, Appl. Phys. Lett., 1:62 (1962).
2. V. S. Bagaev, N. G. Basov, B. M. Vul, B. D. Kopylovskii, O. N. Krokhin, E. P. Markin, Yu. M. Popov, A. N. Khvoshchev, and A. P. Shotov, Dokl. Akad. Nauk SSSR, 150:275 (1963).
3. R. J. Phelan, Jr., and R. H. Rediker, Appl. Phys. Lett., 6:70 (1965).
4. R. L. Bell and K. T. Rogers, Appl. Phys. Lett., 5:9 (1964).
5. A. P. Shotov, S. P. Grishechkina, and R. A. Muminov, Fiz. Tverd. Tela (Leningrad), 8:2496 (1966).

6. G. Dresselhaus, Phys. Rev., 100:580 (1955).
7. R. H. Parmenter, Phys. Rev., 100:573 (1955).
8. E. O. Kane, J. Phys. Chem. Solids, 1:249 (1957).
9. C. Kittel, Quantum Theory of Solids, Wiley, New York (1963).
10. Y. Ohmura, J. Phys. Soc. Jpn., 21:1886 (1966).
11. E. J. Johnson, in: Semiconductors and Semimetals (ed. by R. K. Willardson and A. C. Beer), Vol. 3, Optical Properties of III-V Compounds, Academic Press, New York (1967), p. 154.
12. G. W. Gobeli and H. Y. Fan, Phys. Rev., 119:613 (1960).
13. S. Zwerdling, W. H. Kleiner, and J. P. Theriault, J. Appl. Phys. Suppl., 32:2118 (1961).
14. C. R. Pidgeon and R. N. Brown, Phys. Rev., 146:575 (1966).
15. G. Dresselhaus, J. Phys. Chem. Solids, 1:14 (1956).
16. J. M. Luttinger, Phys. Rev., 102:1030 (1956).
17. G. Dresselhaus and D. M. S. Dresselhaus, Phys. Rev., 160:649 (1967).
18. T. S. Moss, Proc. Phys. Soc. Sect. B, 70:247 (1957).
19. R. J. Phelan, Jr., in: Physics of Quantum Electronics (Proc. Conf. at San Juan, Puerto Rico, 1965), McGraw-Hill, New York (1966), p. 435.
20. J. E. L. Hollis, S. C. Choo, and E. L. Heasell, J. Appl. Phys., 38:1626 (1967).
21. J. Pehek and H. Levinstein, Phys. Rev. 140:A576 (1965).
22. C. Benoit à la Guillaume and P. Lavallard, Proc. Sixth Intern. Conf. on Physics of Semiconductors, Exeter, England, 1962, Institute of Physics, London (1962), p. 875.
23. E. S. Filatova and V. M. Yagodkin, Fiz. Tekh. Poluprovodn., 4:2391 (1970).
24. R. J. Phelan, A. R. Calawa, R. H. Rediker, and R. J. Keyes, and B. Lax, Appl. Phys. Lett., 3:143 (1963).
25. I. Melngailis, R. J. Phelan, and R. H. Rediker, Appl. Phys. Lett., 5:99 (1964).
26. A. P. Shotov, S. P. Grishechkina, B. D. Kopylovskii, and R. A. Muminov, Fiz. Tverd. Tela (Leningrad) 8:1083 (1966).
27. I. I. Zasavitskii, Trudy Fiz. Inst. Akad. Nauk. SSSR, 75:3 (1974).
28. H. Brooks, Adv. Electron. Electron Phys., 7:35 (1955).
29. W. Shockley and J. Bardeen, Phys. Rev., 77:407 (1950).
30. T. P. McLean and E. G. S. Paige, J. Phys. Chem. Solids, 16:220 (1960).
31. J. Appel and R. Bray, Phys. Rev., 127:1603 (1964).
32. H. Schönwald, Z. Naturforsch. Teil A, 19:1276 (1964).
33. H. P. R. Frederikse and W. R. Hosler, Phys. Rev., 108:1146 (1957).
34. G. Fischer, Helv. Phys. Acta, 33:463 (1960).
35. H. J. Hrostowski, F. J. Morin, T. H. Geballe, and G. H. Wheatley, Phys. Rev., 100:1672 (1955).
36. C. H. Champness, J. Electron. Control, 4:201 (1958); Phys. Rev. Lett., 1:439 (1958).
37. G. L. Bir, R. V. Parfen'ev, and P. V. Tamarin, Pis'ma Zh. Eksp. Teor. Fiz., 15:36 (1972).
38. W. D. Straub, W. Bernard, S. Goldstein, and H. Roth, Bull. Am. Phys. Soc., 15:315 (1970).
39. W. Van Roosbroeck and W. Shockley, Phys. Rev., 94:1558 (1954).
40. C. Hilsum, D. J. Oliver, and G. Rickayzen, J. Electron., 1:134 (1955).
41. S. W. Kurnick and R. N. Zitter, J. Appl. Phys., 27:278 (1956).
42. R. N. Zitter, A. J. Strauss, and A. E. Attard, Phys. Rev., 115:266 (1959).
43. V. V. Galavanov, I. A. Kartuzova, and D. N. Nasledov, Fiz. Tverd. Tela Leningrad), 3: 2973 (1961).
44. A. R. Beattie and P. T. Landsberg, Proc. R. Soc. London Ser. A, 249:16 (1959).
45. G. K. Wertheim, Phys. Rev., 104:662 (1956).
46. R. A. Laff and H. Y. Fan, Phys. Rev., 121:53 (1961).
47. D. N. Nasledov, and Yu. S. Smetannikova, Fiz. Tverd. Tela (Leningrad), 4:110 (1962).
48. C. Benoit à la Guillaume and G. Fishman, Phys. Status Solidi, 32:269 (1969).

49. G. Lasher and F. Stern, Phys. Rev., 133:A553 (1964).
50. W. Shockley and W. T. Read, Jr., Phys. Rev., 87:835 (1952).
51. É. K. Guseinov, D. N. Nasledov, A. V. Pentsov, and Yu. G. Popov, Fiz. Tekh. Poluprovodn., 4:179 (1970).
52. A. P. Shotov, and R. A. Muminov, Fiz. Tekh. Poluprovodn., 4:145 (1970).
53. K. Konnerth and C. Lanza, Appl. Phys. Lett., 4:120 (1964).
54. O. Madelung, Physics of III-V Compounds, Wiley, New York (1964).
55. N. G. Basov, B. M. Vul, and Yu. M. Popov, Zh. Eksp. Teor. Fiz., 37:587 (1959).
56. N. G. Basov, O. N. Krokhin, and Yu. M. Popov, Usp. Fiz. Nauk, 72:161 (1960).
57. Yu. M. Popov, Doctoral Thesis, Moscow (1964).
58. N. G. Basov, O. N. Krokhin, and Yu. M. Popov, Zh. Eksp. Teor. Fiz., 40:1879 (1961).
59. É. I. Adirovich and E. M. Kuznetsova, Fiz. Tverd. Tela (Leningrad), 3:3339 (1961).
60. G. Burns and M. I. Nathan, Proc. IEEE, 52:770 (1964).
61. P. G. Eliseev, Thesis for Candidate's Degree, Moscow (1965).
62. A. M. Prokhorov, Zh. Eksp. Teor. Fiz., 34:1658 (1958).
63. A. É. Yunovich, Zh. Tekh. Fiz., 35:1288 (1965).
64. W. P. Dumke, Phys. Rev., 127:1559 (1962).
65. C. Benoit à la Guillaume and P. Lavallard, Solid State Commun., 1:148 (1963).
66. I. Melngailis, Appl. Phys. Lett., 6:59 (1965).
67. H. Welker and H. Weiss, Solid State Phys., 3:1 (1956).
68. A.P. Shotov and S. P. Grishechkina, Fiz. Tverd. Tela (Leningrad), 4:1474 (1962).
69. A. P. Shotov, S. P. Grishechkina, and R. A. Muminov, Zh. Eksp. Teor. Fiz., 50:1525 (1966).
70. B. D. Kopylovskii and V. S. Ivanov, Prib. Tekh. Eksp., No. 4, 145 (1965).
71. F. Herman and L. Petit, Measuring Technique and Electronics [Russian translation], IL, Moscow (1965).
72. D. M. Eagles, J. Phys. Chem. Solids, 16:76 (1960).
73. S. P. Grishechkina and A. P. Shotov, Pis'ma Zh. Eksp. Teor. Fiz., 17:328 (1973).
74. V. L. Bonch-Bruevich and R. Rozman, Fiz. Tverd. Tela (Leningrad), 6:2535 (1964).
75. S. P. Grishechkina and A. P. Shotov, Fiz. Tekh. Poluprovodn., 7:707 (1973).
76. L. V. Keldysh, Zh. Eksp. Teor. Fiz., 34:962 (1958).
77. W. Franz, Z. Naturforsch. Teil A, 13:484 (1958).
78. É. I. Zavaritskaya, Fiz. Tverd. Tela (Leningrad), 2:3009 (1960).
79. V. S. Bagaev, Yu. N. Berozashvili, and L. V. Keldysh, Pis'ma Zh. Eksp. Teor. Fiz., 4:364 (1966).
80. L. M. Roth, B. Lax, and S. Zwerdling, Phys. Rev., 114:90 (1959).
81. Y. Yafet, R. W. Keyes, and E. N. Adams, J. Phys. Chem. Solids, 1:137 (1956).
82. G. Ciobanu, Rev. Roum. Phys., 10:109 (1965).
83. A. P. Shotov and M. S. Murashov, Fiz. Tekh. Poluprovodn., 1:573 (1967).
84. N. A. Penin, B. G. Zhurkin, and B. A. Volkov, Fiz. Tverd. Tela (Leningrad), 7:3188 (1965).
85. B. A. Volkov and V. V. Matveev, Fiz. Tverd. Tela (Leningrad), 8:721 (1966).
86. F. Stern and J. R. Dixon, J. Appl. Phys., 30:268 (1959).
87. S. P. Grishechkina and A. P. Shotov, Kratk. Soobshch. Fiz., No. 7, p. 33 (1973).
88. R. N. Hall, Proc. IRE, 40:1512 (1952).
89. A. A. Lebedev, V. I. Stafeev, and V. M. Tuchkevich, Zh. Tekh. Fiz., 26:2131 (1956).
90. M. A. Lampert, Proc. IRE, 50:1781 (1962).
91. R. H. Lampert and W. Ruppel, J. Appl. Phys., 30:1548 (1959).
92. A. P. Shotov, Doctoral Thesis, Moscow (1967).
93. A. M. Barnett, in: Semiconductors and Semimetals (ed. R. K. Willardson and A. C. Beer), Vol. 6, Injection Phenomena, Academic Press, New York (1970), p. 141.

94. D. A. Kleinman, Bell Syst. Tech. J., 35:685 (1956).
95. G. E. Pikus, Zh. Tekh. Fiz., 27:1606 (1957).
96. G. L. Bir, É. S. Normantas, and G. E. Pikus, Fiz. Tverd. Tela (Leningrad), 4:1180 (1962).
97. G. L. Bir and G. E. Pikus, Fiz. Tverd. Tela (Leningrad), 2:2287 (1960).
98. J. McDougall and E. C. Stoner, Philos. Trans. R. Soc. London Ser. A, 237:67 (1938).